Thermodynamic Tables
to accompany
Modern Engineering
Thermodynamics

Thermodynamic Tables
to accompany
Modern Engineering
Thermodynamics

Robert T. Balmer

AMSTERDAM • BOSTON • HEIDELBERG • LONDON
NEW YORK • OXFORD • PARIS • SAN DIEGO
SAN FRANCISCO • SINGAPORE • SYDNEY • TOKYO
Academic Press is an imprint of Elsevier

Academic Press is an imprint of Elsevier
30 Corporate Drive, Suite 400, Burlington, MA 01803, USA
The Boulevard, Langford Lane, Kidlington, Oxford, OX5 1GB, UK

Library of Congress Cataloging-in-Publication Data
Application submitted

British Library Cataloguing-in-Publication Data
A catalogue record for this book is available from the British Library.

ISBN: 978-0-12-385038-6

For information on all Academic Press publications,
visit our website: *www.elsevierdirect.com*

Typeset by: diacriTech, Chennai, India

Printed in the United States of America
10 11 12 13   6 5 4 3 2 1

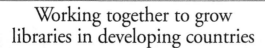

Contents

APPENDIX C Thermodynamic Tables. 1
    C.1a Saturated Water, Temperature Table (English Units). 2
    C.1b Saturated Water, Temperature Table (Metric Units) 4
    C.2a Saturated Water, Pressure Table (English Units) 6
    C.2b Saturated Water, Pressure Table (Metric Units). 8
    C.3a Superheated Water (English Units) 10
    C.3b Superheated Water (Metric Units) 14
    C.4a Compressed Water (English Units). 18
    C.4b Compressed Water (Metric Units) 19
    C.5a Saturated Ammonia (English Units). 21
    C.5b Saturated Ammonia (Metric Units). 22
    C.6a Superheated Ammonia (English Units). 24
    C.6b Superheated Ammonia (Metric Units) 26
    C.7a Saturated Refrigerant-134a, Temperature Table (English Units) 28
    C.7b Saturated Refrigerant-134a, Pressure Table (English Units). 29
    C.7c Saturated Refrigerant-134a, Temperature Table (Metric Units) 30
    C.7d Saturated Refrigerant-134a, Pressure Table (Metric Units). 31
    C.8a Superheated Refrigerant-134a (English Units) 32
    C.8b Superheated Refrigerant-134a (Metric Units) 36
    C.9a Saturated Refrigerant-22, Temperature Table (English Units) 38
    C.9b Saturated Refrigerant-22, Temperature Table (Metric Units) 40
    C.10a Superheated Refrigerant-22 (English Units) 42
    C.10b Superheated Refrigerant-22 (Metric Units). 44
    C.11a Saturated Mercury, Pressure Table (English Units) 46
    C.11b Saturated Mercury, Pressure Table (Metric Units) 47
    C.12a Critical Point Data (English Units) 48
    C.12b Critical Point Data (Metric Units). 49
    C.13a Gas Constant Data (English Units). 50
    C.13b Gas Constant Data (Metric Units) 50
    C.14a Constant Pressure, Specific Heat Ideal Gas Temperature Relations
        (English Units) 51
    C.14b Constant Pressure, Specific Heat Ideal Gas Temperature Relations
        (Metric Units). 51
    C.15a Equation of State Constants (English Units) 52
    C.15b Equation of State Constants (Metric Units) 53
    C.16a Air Tables (English Units) 54
    C.16b Air Tables (Metric Units) 56
    C.16c Other Gases (English Units). 59
    C.17 Base-10 Logarithms of the Equilibrium Constants 61

C.18 Isentropic Compressible Flow Tables for Air ($k = 1.4$) 62
C.19 Normal Shock Tables for Air ($k = 1.4$). 63
C.20 The Elements 64

APPENDIX D Thermodynamic Charts. 65
D.1 *v-u* Chart for Nitrogen 65
D.2 *p-h* Chart for Oxygen. 66
D.3 *T-s* Chart for Water 67
D.4 *T-s* Chart for Carbon Dioxide. 68
D.5 Psychrometric Chart for Water (English Units) 68
D.6 Psychrometric Chart for Water (Metric Units) 69

Thermodynamic Tables

Table	Title	Page
C.1a	Saturated water, temperature table (English units)	2
C.1b	Saturated water, temperature table (metric units)	4
C.2a	Saturated water, pressure table (English units)	6
C.2b	Saturated water, pressure table (metric units)	8
C.3a	Superheated water (English units)	10
C.3b	Superheated water (metric units)	14
C.4a	Compressed water (English units)	18
C.4b	Compressed water (metric units)	19
C.5a	Saturated ammonia (English units)	21
C.5b	Saturated ammonia (metric units)	22
C.6a	Superheated ammonia (English units)	24
C.6b	Superheated ammonia (metric units)	26
C.7a	Saturated Refrigerant-134a, temperature table (English units)	28
C.7b	Saturated Refrigerant-134a, pressure table (English units)	29
C.7c	Saturated Refrigerant-134a, temperature table (metric units)	30
C.7d	Saturated Refrigerant-134a, pressure table (metric units)	31
C.8a	Superheated Refrigerant-134a (English units)	32
C.8b	Superheated Refrigerant-134a (metric units)	36
C.9a	Saturated Refrigerant-22, temperature table (English units)	38
C.9b	Saturated Refrigerant-22, temperature table (metric units)	40
C.10a	Superheated Refrigerant-22 (English units)	42
C.10b	Superheated Refrigerant-22 (metric units)	44
C.11a	Saturated mercury, pressure table (English units)	46
C.11b	Saturated mercury, pressure table (metric units)	47
C.12a	Critical point data (English units)	48
C.12b	Critical point data (metric units)	49
C.13a	Gas constant data (English units)	50
C.13b	Gas constant data (metric units)	50
C.14a	Constant pressure, specific heat ideal gas temperature relations (English units)	51
C.14b	Constant pressure, specific heat ideal gas temperature relations (metric units)	51
C.15a	Equation of state constants (English units)	52
C.15b	Equation of state constants (metric units)	53
C.16a	Air tables (English units)	54
C.16b	Air tables (metric units)	56
C.16c	Other gases (English units)	59
C.17	Base-10 logarithms of the equilibrium constants	61
C.18	Isentropic compressible flow tables for air ($k = 1.4$)	62
C.19	Normal shock tables for air ($k = 1.4$)	63
C.20	The elements	64

Table C.1a Saturated Water, Temperature Table (English Units)

T, °F	p, psia	Volume, ft³/lbm		Energy, Btu/lbm		Enthalpy, Btu/lbm			Entropy, Btu/(lbm · R)		
		v_f	v_g	u_f	u_g	h_f	h_{fg}	h_g	s_f	s_{fg}	s_g
32.018	0.08866	0.01602	3302	0.0	1021.2	0.0	1075.4	1075.4	0.0000	2.1871	2.1871
35	0.09992	0.01602	2948	3.0	1022.2	3.0	1073.7	1076.7	0.0061	2.1705	2.1766
40	0.1217	0.01602	2445	8.0	1023.8	8.0	1070.9	1078.9	0.0162	2.1432	2.1594
45	0.1475	0.01602	2037	13.0	1025.5	13.0	1068.1	1081.1	0.0262	2.1163	2.1425
50	0.1780	0.01602	1704	18.1	1027.2	18.1	1065.2	1083.3	0.0361	2.0900	2.1261
55	0.2140	0.01603	1431	23.1	1028.8	23.1	1062.4	1085.5	0.0458	2.0643	2.1101
60	0.2563	0.01603	1207	28.1	1030.4	28.1	1059.6	1087.7	0.0555	2.0390	2.0945
65	0.3057	0.01604	1021	33.1	1032.1	33.1	1056.8	1089.9	0.0651	2.0142	2.0793
70	0.3632	0.01605	867.6	38.1	1033.7	38.1	1053.9	1092.0	0.0746	1.9898	2.0644
75	0.4300	0.01606	739.7	43.1	1035.4	43.1	1051.1	1094.2	0.0840	1.9659	2.0499
80	0.5073	0.01607	632.7	48.1	1037.0	48.1	1048.3	1096.4	0.0933	1.9425	2.0358
85	0.5964	0.01609	543.1	53.1	1038.6	53.1	1045.5	1098.6	0.1025	1.9195	2.0220
90	0.6989	0.01610	467.6	58.1	1040.2	58.1	1042.6	1100.7	0.1116	1.8969	2.0085
95	0.8162	0.01611	403.9	63.0	1041.9	63.1	1039.8	1102.9	0.1206	1.8747	1.9953
100	0.9503	0.01613	350.0	68.0	1043.5	68.0	1037.0	1105.0	0.1296	1.8528	1.9824
110	1.276	0.01617	265.1	78.0	1046.7	78.0	1031.3	1109.3	0.1473	1.8103	1.9576
120	1.695	0.01621	203.0	88.0	1049.9	88.0	1025.5	1113.5	0.1646	1.7692	1.9338
130	2.225	0.01625	157.2	98.0	1053.0	98.0	1019.7	1117.7	0.1817	1.7294	1.9111
140	2.892	0.01629	122.9	107.9	1056.2	108.0	1013.9	1121.9	0.1985	1.6909	1.8894
150	3.722	0.01634	96.98	117.9	1059.3	117.9	1008.2	1126.1	0.2150	1.6535	1.8685
160	4.745	0.01640	77.23	127.9	1062.3	128.0	1002.1	1130.1	0.2313	1.6173	1.8486
180	7.515	0.01651	50.20	148.0	1068.3	148.0	990.2	1138.2	0.2631	1.5480	1.8111
200	11.53	0.01663	33.63	168.0	1074.2	168.1	977.8	1145.9	0.2941	1.4823	1.7764
212	14.696	0.01672	26.80	180.1	1077.6	180.1	970.4	1150.5	0.3122	1.4447	1.7569
220	17.19	0.01677	23.15	188.2	1079.8	188.2	965.3	1153.5	0.3241	1.4202	1.7443
240	24.97	0.01692	16.33	208.4	1085.3	208.4	952.3	1160.7	0.3534	1.3611	1.7145
260	35.42	0.01708	11.77	228.6	1090.5	228.7	938.9	1167.6	0.3820	1.3046	1.6866

T, °F	p, psia	Volume, ft³/lbm		Energy, Btu/lbm		Enthalpy, Btu/lbm			Entropy, Btu/(lbm · R)		
		v_f	v_g	u_f	u_g	h_f	h_{fg}	h_g	s_f	s_{fg}	s_g
280	49.19	0.01726	8.650	249.0	1095.4	249.2	924.9	1174.1	0.4100	1.2504	1.6604
300	66.98	0.01745	6.472	269.5	1100.0	269.7	910.5	1180.2	0.4373	1.1985	1.6358
320	89.60	0.01765	4.919	290.1	1104.2	290.4	895.4	1185.8	0.4641	1.1484	1.6125
340	117.9	0.01787	3.792	310.9	1108.0	311.3	879.5	1190.8	0.4904	1.0998	1.5902
360	152.9	0.01811	2.961	331.8	1111.4	332.3	862.9	1195.2	0.5163	1.0527	1.5690
380	195.6	0.01836	2.339	352.9	1114.3	353.6	845.4	1199.0	0.5417	1.0068	1.5485
400	247.1	0.01864	1.866	374.3	1116.6	375.1	826.9	1202.0	0.5668	0.9618	1.5286
420	308.5	0.01894	1.502	395.8	1118.3	396.9	807.2	1204.1	0.5916	0.9177	1.5093
440	381.2	0.01926	1.219	417.6	1119.3	419.0	786.3	1205.3	0.6162	0.8740	1.4902
460	466.3	0.01961	0.9961	439.7	1119.6	441.4	764.1	1205.5	0.6405	0.8309	1.4714
480	565.5	0.02000	0.8187	462.2	1118.9	464.3	740.3	1204.6	0.6647	0.7879	1.4526
500	680.0	0.02043	0.6761	485.1	1117.4	487.7	714.8	1202.5	0.6889	0.7448	1.4337
520	811.4	0.02091	0.5605	508.5	1114.8	511.7	687.2	1198.9	0.7132	0.7015	1.4147
540	961.4	0.02145	0.4658	532.6	1111.0	536.4	657.4	1193.8	0.7375	0.6577	1.3952
560	1132	0.02207	0.3877	557.3	1105.8	562.0	625.0	1187.0	0.7622	0.6129	1.3751
580	1324	0.02278	0.3225	583.0	1098.9	588.6	589.4	1178.0	0.7873	0.5669	1.3542
600	1541	0.02363	0.2677	609.9	1090.0	616.6	549.8	1166.4	0.8131	0.5188	1.3319
620	1784	0.02465	0.2209	638.3	1078.5	646.4	505.0	1151.4	0.8399	0.4678	1.3077
640	2057	0.02593	0.1805	668.7	1063.2	678.6	453.3	1131.9	0.8683	0.4122	1.2805
660	2362	0.02767	0.1446	702.0	1042.3	714.3	391.2	1105.5	0.8992	0.3493	1.2485
680	2705	0.03032	0.1113	741.7	1011.0	756.9	309.8	1066.7	0.9352	0.2718	1.2070
700	3090	0.03666	0.07444	801.7	947.7	822.7	167.7	990.4	0.9903	0.1447	1.1350
705.445	3203.8	0.05053	0.05053	872.6	872.6	902.5	0.0	902.5	1.0582	0.0000	1.0582

Note: Saturated liquid entropies have been adjusted to make the Gibbs functions of the liquid and vapor phases exactly equal. For this reason, there are some small differences between values presented here and the original tables.

Sources: Reprinted by permission from Reynolds, W. C., Perkins, H. C. Engineering Thermodynamics, second ed., 1977, McGraw-Hill, New York. Recalculated from equations given in Keenan, J. H., Keyes, F. G., Hill, P. G., Moore, J. G. Steam Tables. Wiley, New York, 1969. Reprinted by permission of John Wiley & Sons, Inc.

Table C.1b Saturated Water, Temperature Table (Metric Units)

| T, °C | p, MPa | Volume, m³/kg | | Energy, kJ/kg | | | Enthalpy, kJ/kg | | | Entropy, kJ/(kg · K) | | |
		v_f	v_g	u_f	u_g		h_f	h_{fg}	h_g	s_f	s_{fg}	s_g
0.010	0.0006113	0.001000	206.1	0.0	2375.3		0.0	2501.3	2501.3	0.0000	9.1571	9.1571
2	0.0007056	0.001000	179.9	8.4	2378.1		8.4	2496.6	2505.0	0.0305	9.0738	9.1043
5	0.0008721	0.001000	147.1	21.0	2382.2		21.0	2489.5	2510.5	0.0761	8.9505	9.0266
10	0.001228	0.001000	106.4	42.0	2389.2		42.0	2477.7	2519.7	0.1510	8.7506	8.9016
15	0.001705	0.001001	77.93	63.0	2396.0		63.0	2465.9	2528.9	0.2244	8.5578	8.7822
20	0.002338	0.001002	57.79	83.9	2402.9		83.9	2454.2	2538.1	0.2965	8.3715	8.6680
25	0.003169	0.001003	43.36	104.9	2409.8		104.9	2442.3	2547.2	0.3672	8.1916	8.5588
30	0.004246	0.001004	32.90	125.8	2416.6		125.8	2430.4	2556.2	0.4367	8.0174	8.4541
35	0.005628	0.001006	25.22	146.7	2423.4		146.7	2418.6	2565.3	0.5051	7.8488	8.3539
40	0.007383	0.001008	19.52	167.5	2430.1		167.5	2406.8	2574.3	0.5723	7.6855	8.2578
45	0.009593	0.001010	15.26	188.4	2436.8		188.4	2394.8	2583.2	0.6385	7.5271	8.1656
50	0.01235	0.001012	12.03	209.3	2443.5		209.3	2382.8	2592.1	0.7036	7.3735	8.0771
55	0.01576	0.001015	9.569	230.2	2450.1		230.2	2370.7	2600.9	0.7678	7.2243	7.9921
60	0.01994	0.001017	7.671	251.1	2456.6		251.1	2358.5	2609.6	0.8310	7.0794	7.9104
65	0.02503	0.001020	6.197	272.0	2463.1		272.0	2346.2	2618.2	0.8934	6.9384	7.8318
70	0.03119	0.001023	5.042	292.9	2469.5		293.0	2333.8	2626.8	0.9549	6.8012	7.7561
75	0.03858	0.001026	4.131	313.9	2475.9		313.9	2321.4	2635.3	1.0155	6.6678	7.6833
80	0.04739	0.001029	3.407	334.8	2482.2		334.9	2308.8	2643.7	1.0754	6.5376	7.6130
85	0.05783	0.001032	2.828	355.8	2488.4		355.9	2296.0	2651.9	1.1344	6.4109	7.5453
90	0.07013	0.001036	2.361	376.8	2494.5		376.9	2283.2	2660.1	1.1927	6.2872	7.4799
95	0.08455	0.001040	1.982	397.9	2500.6		397.9	2270.2	2668.1	1.2503	6.1664	7.4167
100	0.1013	0.001044	1.673	418.9	2506.5		419.0	2257.0	2676.0	1.3071	6.0486	7.3557
110	0.1433	0.001052	1.210	461.1	2518.1		461.3	2230.2	2691.5	1.4188	5.8207	7.2395
120	0.1985	0.001060	0.8919	503.5	2529.2		503.7	2202.6	2706.3	1.5280	5.6024	7.1304
130	0.2701	0.001070	0.6685	546.0	2539.9		546.3	2174.2	2720.5	1.6348	5.3929	7.0277
140	0.3613	0.001080	0.5089	588.7	2550.0		589.1	2144.8	2733.9	1.7395	5.1912	6.9307
150	0.4758	0.001090	0.3928	631.7	2559.5		632.2	2114.2	2746.4	1.8422	4.9965	6.8387

T, °C	p, MPa	Volume, m³/kg		Energy, kJ/kg		Enthalpy, kJ/kg			Entropy, kJ/(kg·K)		
		v_f	v_g	u_f	u_g	h_f	h_{fg}	h_g	s_f	s_{fg}	s_g
160	0.6178	0.001102	0.3071	674.9	2568.4	675.5	2082.6	2758.1	1.9431	4.8079	6.7510
170	0.7916	0.001114	0.2428	718.3	2576.5	719.2	2049.5	2768.7	2.0423	4.6249	6.6672
180	1.002	0.001127	0.1941	762.1	2583.7	763.2	2015.0	2778.2	2.1400	4.4466	6.5866
190	1.254	0.001141	0.1565	806.2	2590.0	807.5	1978.8	2786.4	2.2363	4.2724	6.5087
200	1.554	0.001156	0.1274	850.6	2595.3	852.4	1940.8	2793.2	2.3313	4.1018	6.4331
210	1.906	0.001173	0.1044	895.5	2599.4	897.7	1900.8	2798.5	2.4253	3.9340	6.3593
220	2.318	0.001190	0.08620	940.9	2602.4	943.6	1858.5	2802.1	2.5183	3.7686	6.2869
230	2.795	0.001209	0.07159	986.7	2603.9	990.1	1813.9	2804.0	2.6105	3.6050	6.2155
240	3.344	0.001229	0.05977	1033.2	2604.0	1037.3	1766.5	2803.8	2.7021	3.4425	6.1446
250	3.973	0.001251	0.05013	1080.4	2602.4	1085.3	1716.2	2801.5	2.7933	3.2805	6.0738
260	4.688	0.001276	0.04221	1128.4	2599.0	1134.4	1662.5	2796.9	2.8844	3.1184	6.0028
270	5.498	0.001302	0.03565	1177.3	2593.7	1184.5	1605.2	2789.7	2.9757	2.9553	5.9310
280	6.411	0.001332	0.03017	1227.4	2586.1	1236.0	1543.6	2779.6	3.0674	2.7905	5.8579
290	7.436	0.001366	0.02557	1278.9	2576.0	1289.0	1477.2	2766.2	3.1600	2.6230	5.7830
300	8.580	0.001404	0.02168	1332.0	2563.0	1344.0	1405.0	2749.0	3.2540	2.4513	5.7053
310	9.856	0.001447	0.01835	1387.0	2546.4	1401.3	1326.0	2727.3	3.3500	2.2739	5.6239
320	11.27	0.001499	0.01549	1444.6	2525.5	1461.4	1238.7	2700.1	3.4487	2.0883	5.5370
330	12.84	0.001561	0.01300	1505.2	2499.0	1525.3	1140.6	2665.9	3.5514	1.8911	5.4425
340	14.59	0.001638	0.01080	1570.3	2464.6	1594.2	1027.9	2622.1	3.6601	1.6765	5.3366
350	16.51	0.001740	0.008815	1641.8	2418.5	1670.6	893.4	2564.0	3.7784	1.4338	5.2122
360	18.65	0.001892	0.006947	1725.2	2351.6	1760.5	720.7	2481.2	3.9154	1.1382	5.0536
370	21.03	0.002213	0.004931	1844.0	2229.0	1890.5	442.2	2332.7	4.1114	0.6876	4.7990
374.136	22.088	0.003155	0.003155	2029.6	2029.6	2099.3	0.0	2099.3	4.4305	0.0000	4.4305

Note: Saturated liquid entropies have been adjusted to make the Gibbs functions of the liquid and vapor phases exactly equal. For this reason, there are some small differences between values presented here and the original tables.

Sources: Reprinted by permission from Reynolds, W. C., Perkins, H. C. Engineering Thermodynamics, second ed., 1977, McGraw-Hill, New York. Recalculated from equations given in Keenan, J. H., Keyes, F. G., Hill, P. G., Moore, J. G. Steam Tables. Wiley, New York, 1969. Reprinted by permission of John Wiley & Sons, Inc.

Table C.2a Saturated Water, Pressure Table (English Units)

| p, psia | T, °F | Volume, ft³/lbm | | Energy, Btu/lbm | | | Enthaply, Btu/lbm | | | Entropy, Btu/(lbm · R) | | |
		v_f	v_g	u_f	u_g	h_f	h_{fg}	h_g	s_f	s_{fg}	s_g
0.0887	32.018	0.01602	3302.0	0.0	1021.2	0.0	1075.4	1075.4	0.0000	2.1871	2.1871
0.1	35.0	0.01602	2946.0	3.0	1022.2	3.0	1073.7	1076.7	0.0061	2.1705	2.1768
0.12	39.6	0.01602	2477.0	7.7	1023.7	7.7	1071.0	1078.7	0.0155	2.1451	2.1606
0.14	43.6	0.01602	2140.0	11.7	1025.0	11.7	1068.8	1080.5	0.0234	2.1237	2.1471
0.16	47.1	0.01602	1886.0	15.2	1026.2	15.2	1066.8	1082.0	0.0304	2.1050	2.1354
0.18	50.3	0.01602	1686.0	18.3	1027.2	18.3	1065.1	1083.4	0.0366	2.0886	2.1252
0.2	53.1	0.01603	1526.0	21.2	1028.2	21.2	1063.5	1084.7	0.0422	2.0738	2.1160
0.25	59.3	0.01603	1235.0	27.4	1030.2	27.4	1060.0	1087.4	0.0542	2.0425	2.0967
0.3	64.5	0.01604	1040.0	32.5	1031.9	32.5	1057.1	1089.6	0.0641	2.0168	2.0809
0.4	72.8	0.01606	792.0	40.9	1034.7	40.9	1052.4	1093.3	0.0799	1.9762	2.0561
0.6	85.2	0.01609	540.0	53.3	1038.7	53.3	1045.3	1098.6	0.1028	1.9187	2.0215
0.8	94.3	0.01611	411.7	62.4	1041.6	62.4	1040.2	1102.6	0.1195	1.8775	1.9970
1	101.7	0.01614	333.6	69.7	1044.0	69.7	1036.0	1105.7	0.1326	1.8455	1.9781
1.2	107.9	0.01616	280.9	75.9	1046.0	75.9	1032.5	1108.4	0.1435	1.8193	1.9628
1.4	113.2	0.01618	243.0	81.2	1047.7	81.2	1029.5	1110.7	0.1529	1.7969	1.9498
1.6	117.9	0.01620	214.3	85.9	1049.2	85.9	1026.8	1112.7	0.1611	1.7775	1.9386
1.8	122.2	0.01621	191.8	90.2	1050.6	90.2	1024.3	1114.5	0.1684	1.7604	1.9288
2	126.0	0.01623	173.8	94.0	1051.8	94.0	1022.1	1116.1	0.1750	1.7450	1.9200
3	141.4	0.01630	118.7	109.4	1056.6	109.4	1013.1	1122.5	0.2009	1.6854	1.8863
4	152.9	0.01636	90.64	120.9	1060.2	120.9	1006.4	1127.3	0.2198	1.6428	1.8626
6	170.0	0.01645	61.98	138.0	1065.4	138.0	996.2	1134.2	0.2474	1.5820	1.8294
8	182.8	0.01653	47.35	150.8	1069.2	150.8	988.5	1139.3	0.2676	1.5384	1.8060
10	193.2	0.01659	38.42	161.2	1072.2	161.2	982.1	1143.3	0.2836	1.5043	1.7879
12	201.9	0.01665	32.40	170.0	1074.7	170.0	976.7	1146.7	0.2970	1.4762	1.7732
14	209.6	0.01670	28.05	177.6	1076.9	177.7	971.9	1149.6	0.3085	1.4523	1.7608
14.696	212.0	0.01672	26.80	180.1	1077.6	180.1	970.4	1150.5	0.3122	1.4447	1.7569
16	216.3	0.01675	24.75	184.4	1078.8	184.5	967.6	1152.1	0.3186	1.4315	1.7501

p, psia	T, °F	Volume, ft³/lbm		Energy, Btu/lbm		Enthaply, Btu/lbm			Ertropy, Btu/(lbm · R)		
		v_f	v_g	u_f	u_g	h_f	h_{fg}	h_g	s_f	s_{fg}	s_g
18	222.4	0.01679	22.17	190.6	1080.5	190.6	963.8	1154.4	0.3277	1.4129	1.7406
20	228.0	0.01683	20.09	196.2	1082.0	196.2	960.2	1156.4	0.3359	1.3963	1.7322
30	250.3	0.01700	13.75	218.8	1088.0	218.9	945.4	1164.3	0.3683	1.3315	1.6998
40	267.3	0.01715	10.50	236.0	1092.3	236.2	933.8	1170.0	0.3922	1.2847	1.6769
60	292.7	0.01738	7.177	262.0	1098.3	262.2	915.8	1178.0	0.4274	1.2172	1.6446
80	312.1	0.01757	5.474	281.9	1102.6	282.2	901.4	1183.6	0.4535	1.1681	1.6216
100	327.9	0.01774	4.434	298.3	1105.8	298.6	889.2	1187.8	0.4745	1.1291	1.6036
120	341.3	0.01789	3.730	312.3	1108.3	312.7	878.4	1191.1	0.4921	1.0967	1.5888
140	353.1	0.01802	3.221	324.6	1110.3	325.0	868.8	1193.8	0.5074	1.0688	1.5762
160	363.6	0.01815	2.836	335.6	1112.0	336.2	859.8	1196.0	0.5209	1.0443	1.5652
180	373.1	0.01827	2.533	345.7	1113.4	346.3	851.5	1197.8	0.5330	1.0225	1.5555
200	381.9	0.01839	2.289	354.9	1114.6	355.6	843.7	1199.3	0.5441	1.0025	1.5466
300	417.4	0.01890	1.544	393.0	1118.1	394.1	809.8	1203.9	0.5885	0.9232	1.5117
400	444.7	0.01934	1.162	422.8	1119.4	424.2	781.3	1205.5	0.6219	0.8639	1.4858
600	486.3	0.02013	0.7702	469.4	1118.5	471.6	732.5	1204.1	0.6724	0.7742	1.4466
800	518.4	0.02087	0.5691	506.6	1115.0	509.7	689.6	1199.3	0.7112	0.7050	1.4162
1000	544.8	0.02159	0.4459	538.4	1109.9	542.4	650.0	1192.4	0.7434	0.6471	1.3905
1200	567.4	0.02232	0.3623	566.7	1103.5	571.7	612.2	1183.9	0.7714	0.5961	1.3675
1400	587.3	0.02307	0.3016	592.6	1096.0	598.6	575.5	1174.1	0.7966	0.5497	1.3463
1600	605.1	0.02386	0.2552	616.9	1087.4	624.0	538.9	1162.9	0.8198	0.5062	1.3260
2000	636.0	0.02565	0.1881	662.4	1066.6	671.9	464.4	1136.3	0.8624	0.4239	1.2863
2600	674.1	0.02938	0.1210	729.2	1021.8	743.3	336.8	1080.1	0.9237	0.2971	1.2208
3203.8	705.445	0.05053	0.05053	872.6	872.6	902.5	0.0	902.5	1.0582	0.0000	1.0582

Note: Saturated liquid entropies have been adjusted to make the Gibbs functions of the liquid and vapor phases exactly equal. For this reason, there are some small differences between values presented here and the original tables.

Sources: Reprinted by permission from Reynolds, W. C., Perkins, H. C. Engineering Thermodynamics, second ed., 1977, McGraw-Hill, New York. Recalculated from equations given in Keenan, J. H., Keyes, F. G., Hill, P. G., Moore, J. G. Steam Tables. Wiley, New York, 1969. Reprinted by permission of John Wiley & Sons, Inc.

Table C.2b Saturated Water, Pressure Table (Metric Units)

p, Mpa	T, °C	Volume, m³/kg		Energy, kJ/kg			Enthaply, kJ/kg			Entropy, kJ/(kg·K)		
		v_f	v_g	u_f	u_g	h_f	h_{fg}	h_g	s_f	s_{fg}	s_g
0.000611	0.01	0.001000	206.1	0.0	2375.3	0.0	2501.3	2501.3	0.0000	9.1571	9.1571
0.0008	3.8	0.001000	159.7	15.8	2380.5	15.8	2492.5	2508.3	0.0575	9.0007	9.0582
0.001	7.0	0.001000	129.2	29.3	2385.0	29.3	2484.9	2514.2	0.1059	8.8706	8.9765
0.0012	9.7	0.001000	108.7	40.6	2388.7	40.6	2478.5	2519.1	0.1460	8.7639	8.9099
0.0014	12.0	0.001001	93.92	50.3	2391.9	50.3	2473.1	2523.4	0.1802	8.6736	8.8538
0.0016	14.0	0.001001	82.76	58.9	2394.7	58.9	2468.2	2527.1	0.2101	8.5952	8.8053
0.0018	15.8	0.001001	74.03	66.5	2397.2	66.5	2464.0	2530.5	0.2367	8.5259	8.7626
0.002	17.5	0.001001	67.00	73.5	2399.5	73.5	2460.0	2533.5	0.2606	8.4639	8.7245
0.003	24.1	0.001003	45.67	101.0	2408.5	101.0	2444.5	2545.5	0.3544	8.2240	8.5784
0.004	29.0	0.001004	34.80	121.4	2415.2	121.4	2433.0	2554.4	0.4225	8.0529	8.4754
0.006	36.2	0.001006	23.74	151.5	2424.9	151.5	2415.9	2567.4	0.5208	7.8104	8.3312
0.008	41.5	0.001008	18.10	173.9	2432.1	173.9	2403.1	2577.0	0.5924	7.6371	8.2295
0.01	45.8	0.001010	14.67	191.8	2437.9	191.8	2392.8	2584.6	0.6491	7.5019	8.1510
0.012	49.4	0.001012	12.36	206.9	2442.7	206.9	2384.1	2591.0	0.6961	7.3910	8.0871
0.014	52.6	0.001013	10.69	220.0	2446.9	220.0	2376.6	2596.6	0.7365	7.2968	8.0333
0.016	55.3	0.001015	9.433	231.5	2450.5	231.5	2369.9	2601.4	0.7719	7.2149	7.9868
0.018	57.8	0.001016	8.445	241.9	2453.8	241.9	2363.9	2605.8	0.8034	7.1425	7.9459
0.02	60.1	0.001017	7.649	251.4	2456.7	251.4	2358.3	2609.7	0.8319	7.0774	7.9093
0.03	69.1	0.001022	5.229	289.2	2468.4	289.2	2336.1	2625.3	0.9439	6.8256	7.7695
0.04	75.9	0.001026	3.993	317.5	2477.0	317.6	2319.1	2636.7	1.0260	6.6449	7.6709
0.06	85.9	0.001033	2.732	359.8	2489.6	359.8	2293.7	2653.5	1.1455	6.3873	7.5328
0.08	93.5	0.001039	2.087	391.6	2498.8	391.6	2274.1	2665.7	1.2331	6.2023	7.4354
0.1	99.6	0.001043	1.694	417.3	2506.1	417.4	2258.1	2675.5	1.3029	6.0573	7.3602
0.12	104.8	0.001047	1.428	439.2	2512.1	439.3	2244.2	2683.5	1.3611	5.9378	7.2989
0.14	109.3	0.001051	1.237	458.2	2517.3	458.4	2232.0	2690.4	1.4112	5.8360	7.2472
0.16	113.3	0.001054	1.091	475.2	2521.8	475.3	2221.2	2696.5	1.4553	5.7472	7.2025
0.18	116.9	0.001058	0.9775	490.5	2525.9	490.7	2211.1	2701.8	1.4948	5.6683	7.1631

p, Mpa	T, °C	Volume, m³/kg		Energy, kJ/kg		Enthaply, kJ/kg			Entropy, kJ/(kg · K)		
		v_f	v_g	u_f	u_g	h_f	h_{fg}	h_g	s_f	s_{fg}	s_g
0.2	120.2	0.001061	0.8857	504.5	2529.5	504.7	2201.9	2706.6	1.5305	5.5975	7.1280
0.3	133.5	0.001073	0.6058	561.1	2543.6	561.5	2163.8	2725.3	1.6722	5.3205	6.9927
0.4	143.6	0.001084	0.4625	604.3	2553.6	604.7	2133.8	2738.5	1.7770	5.1197	6.8967
0.6	158.9	0.001101	0.3157	669.9	2567.4	670.6	2086.2	2756.8	1.9316	4.8293	6.7609
0.8	170.4	0.001115	0.2404	720.2	2576.8	721.1	2048.0	2769.1	2.0466	4.6170	6.6636
1	179.9	0.001127	0.1944	761.7	2583.6	762.8	2015.3	2778.1	2.1391	4.4482	6.5873
1.2	188.0	0.001139	0.1633	797.3	2588.8	798.6	1986.2	2784.8	2.2170	4.3072	6.5242
1.4	195.1	0.001149	0.1408	828.7	2592.8	830.3	1959.7	2790.0	2.2847	4.1854	6.4701
1.6	201.4	0.001159	0.1238	856.9	2596.0	858.8	1935.2	2794.0	2.3446	4.0780	6.4226
1.8	207.2	0.001168	0.1104	882.7	2598.4	884.8	1912.3	2797.1	2.3986	3.9816	6.3802
2	212.4	0.001177	0.09963	906.4	2600.3	908.8	1890.7	2799.5	2.4478	3.8939	6.3417
3	233.9	0.001216	0.06668	1004.8	2604.1	1008.4	1795.7	2804.1	2.6462	3.5416	6.1878
4	250.4	0.001252	0.04978	1082.3	2602.3	1087.3	1714.1	2801.4	2.7970	3.2739	6.0709
6	275.6	0.001319	0.03244	1205.4	2589.7	1213.3	1571.0	2784.3	3.0273	2.8627	5.8900
8	295.1	0.001384	0.02352	1305.6	2569.8	1316.6	1441.4	2758.0	3.2075	2.5365	5.7440
9	303.4	0.001418	0.02048	1350.5	2557.8	1363.3	1378.8	2742.1	3.2865	2.3916	5.6781
10	311.1	0.001452	0.01803	1393.0	2544.4	1407.6	1317.1	2724.7	3.3603	2.2546	5.6149
12	324.8	0.001527	0.01426	1472.9	2513.7	1491.3	1193.6	2684.9	3.4970	1.9963	5.4933
14	336.8	0.001611	0.01149	1548.6	2476.8	1571.1	1066.5	2637.6	3.6240	1.7486	5.3726
16	347.4	0.001711	0.009307	1622.7	2431.8	1650.0	930.7	2580.7	3.7468	1.4996	5.2464
18	357.1	0.001840	0.007491	1698.9	2374.4	1732.0	777.2	2509.2	3.8722	1.2332	5.1054
20	365.8	0.002036	0.005836	1785.6	2293.2	1826.3	583.7	2410.0	4.0146	0.9135	4.9281
22.088	374.136	0.003155	0.003155	2029.6	2029.6	2099.3	0.0	2099.3	4.4305	0.0000	4.4305

Note: Saturated liquid entropies have been adjusted to make the Gibbs functions of the liquid and vapor phases exactly equal. For this reason, there are some small differences between values presented here and the original tables.

Sources: Reprinted by permission from Reynolds, W. C., Perkins, H. C. Engineering Thermodynamics, second ed., 1977, McGraw-Hill, New York. Recalculated from equations given in Keenan, J. H., Keyes, F. G., Hill, P. G., Moore, J. G. Steam Tables. Wiley, New York, 1969. Reprinted by permission of John Wiley & Sons, Inc.

Table C.3a Superheated Water Vapor (English Units)

p, psia (T_sat, °F)		Temperature, °F												
		100	200	300	400	500	600	700	800	900	1000	1100	1200	1300
0.2 (53.1)	v, ft³/lbm	1666.0	1964.0	2262.0	2560.0	2858.0	3156.0	3454.0	3752.0	4050.0	4347.0	4645.0	4943.0	5241.0
	u, Btu/lbm	1043.9	1077.7	1112.1	1147.1	1182.8	1219.3	1256.7	1294.9	1334.0	1373.9	1414.8	1456.7	1499.4
	h, Btu/lbm	1105.6	1150.4	1195.8	1241.9	1288.6	1336.1	1384.5	1433.7	1483.8	1534.8	1586.8	1639.6	1693.4
	s, Btu/(lbm·R)	2.1550	2.2287	2.2928	2.3497	2.4011	2.4483	2.4919	2.5326	2.5708	2.607	2.6414	2.6743	2.7058
0.5 (79.5)	v, ft³/lbm	665.9	785.5	904.8	1024.0	1143.0	1262.0	1382.0	1501.0	1620.0	1739.0	1858.0	1977.0	2096.0
	u, Btu/lbm	1043.7	1077.6	1112.0	1147.1	1182.8	1219.3	1256.7	1294.9	1334.0	1373.9	1414.8	1456.7	1499.4
	h, Btu/lbm	1105.3	1150.3	1195.8	1241.8	1288.6	1336.1	1384.5	1433.7	1483.8	1534.8	1586.8	1639.6	1693.4
	s, Btu/(lbm·R)	2.0537	2.1276	2.1917	2.2487	2.3001	2.3472	2.3909	2.4316	2.4698	2.506	2.5404	2.5733	2.6048
1 (101.7)	v, ft³/lbm	—	392.5	452.3	511.9	571.5	631.1	690.7	750.3	809.9	869.5	929.0	988.6	1048.0
	u, Btu/lbm	—	1077.5	1112.0	1147.0	1182.8	1219.3	1256.7	1294.9	1333.9	1373.9	1414.8	1456.7	1499.4
	h, Btu/lbm	—	1150.1	1195.6	1241.7	1288.5	1336.1	1384.5	1433.7	1483.8	1534.8	1586.8	1639.6	1693.4
	s, Btu/(lbm·R)	—	2.0510	2.1152	2.1722	2.2237	2.2708	2.3144	2.3551	2.3934	2.4296	2.4640	2.4969	2.5283
2 (126.0)	v, ft³/lbm	—	196.0	226.0	255.9	285.7	315.5	345.3	375.1	404.9	434.7	464.5	494.3	524.1
	u, Btu/lbm	—	1077.2	1111.8	1146.9	1182.7	1219.3	1256.6	1294.8	1333.9	1373.9	1414.8	1456.7	1499.4
	h, Btu/lbm	—	1149.7	1195.4	1241.6	1288.4	1336.0	1384.4	1433.7	1483.8	1534.8	1586.7	1639.6	1693.4
	s, Btu/(lbm·R)	—	1.9741	2.0386	2.0957	2.1472	2.1944	2.2380	2.2787	2.3170	2.3532	2.3876	2.4205	2.4519
5 (162.2)	v, ft³/lbm	—	78.15	90.24	102.2	114.2	126.1	138.1	150.0	161.9	173.9	185.8	197.7	209.6
	u, Btu/lbm	—	1076.2	1111.3	1146.6	1182.5	1219.1	1256.5	1294.7	1333.8	1373.8	1414.8	1456.6	1499.4
	h, Btu/lbm	—	1148.6	1194.8	1241.2	1288.2	1335.8	1384.3	1433.5	1483.7	1534.7	1586.7	1639.5	1693.3
	s, Btu/(lbm·R)	—	1.8717	1.9369	1.9943	2.0460	2.0932	2.1369	2.1776	2.2159	2.2522	2.2866	2.3194	2.3509
10 (193.2)	v, ft³/lbm	—	38.85	44.99	51.03	57.04	63.03	69.01	74.98	80.95	86.91	92.88	98.84	104.8
	u, Btu/lbm	—	1074.7	1110.4	1146.1	1182.2	1218.8	1256.3	1294.6	1333.7	1373.7	1414.7	1456.5	1499.3
	h, Btu/lbm	—	1146.6	1193.7	1240.5	1287.7	1335.5	1384.0	1433.3	1483.5	1534.6	1586.5	1639.4	1693.2
	s, Btu/(lbm·R)	—	1.7929	1.8594	1.9173	1.9692	2.0166	2.0603	2.1011	2.1394	2.1757	2.2101	2.2430	2.2745
14.7 (212.0)	v, ft³/lbm	—	—	30.52	34.67	38.77	42.86	46.93	51.00	55.07	59.13	63.19	67.25	71.30
	u, Btu/lbm	—	—	1109.6	1145.6	1181.8	1218.6	1256.1	1294.4	1333.6	1373.6	1414.6	1456.5	1499.3
	h, Btu/lbm	—	—	1192.6	1239.9	1287.3	1335.2	1383.8	1433.1	1483.3	1534.4	1586.4	1639.3	1693.2
	s, Btu/(lbm·R)	—	—	1.8159	1.8743	1.9264	1.9739	2.0177	2.0586	2.0969	2.1332	2.1676	2.2005	2.2320

Temperature, °F

p, psia (T_sat, °F)		300	400	500	600	700	800	900	1000	1100	1200	1300	1400	1500
20 (227.9)	v, ft³/lbm	22.36	25.43	28.46	31.47	34.47	37.46	40.45	43.44	46.42	49.41	52.39	55.37	58.35
	u, Btu/lbm	1108.7	1145.1	1181.5	1218.3	1255.9	1294.3	1333.5	1373.5	1414.5	1456.4	1499.2	1542.9	1587.6
	h, Btu/lbm	1191.4	1239.2	1286.8	1334.8	1383.5	1432.2	1483.3	1534.3	1586.3	1639.2	1693.1	1747.8	1803.5
	s, Btu/(lbm·R)	1.7807	1.8397	1.8921	1.9397	1.9836	2.0245	2.0629	2.0991	2.1336	2.1665	2.1980	2.2283	2.2574
40 (267.2)	v, ft³/lbm	11.04	12.62	14.16	15.69	17.20	18.70	20.20	21.70	23.20	24.69	26.18	27.68	29.17
	u, Btu/lbm	1105.1	1143.0	1180.1	1217.3	1255.1	1293.7	1333.0	1373.1	1414.2	1456.1	1498.9	1542.7	1587.4
	h, Btu/lbm	1186.8	1236.4	1284.9	1333.4	1382.4	1432.1	1482.5	1533.7	1585.9	1638.9	1692.8	1747.6	1803.3
	s, Btu/(lbm·R)	1.6995	1.7608	1.8142	1.8623	1.9065	1.9476	1.9861	2.0224	2.0570	2.0899	2.1214	2.1517	2.1809
60 (292.7)	v, ft³/lbm	7.260	8.353	9.399	10.42	11.44	12.45	13.45	14.45	15.45	16.45	17.45	18.45	19.44
	u, Btu/lbm	1101.3	1140.8	1178.6	1216.3	1254.4	1293.0	1332.5	1372.7	1413.8	1455.8	1498.7	1542.5	1587.2
	h, Btu/lbm	1181.9	1233.5	1283.0	1332.1	1381.4	1431.2	1481.8	1533.2	1585.4	1638.5	1692.4	1747.3	1803.0
	s, Btu/(lbm·R)	1.6497	1.7136	1.7680	1.8167	1.8611	1.9024	1.9410	1.9775	2.0121	2.0450	2.0766	2.1069	2.1361
80 (312.0)	v, ft³/lbm	—	6.217	7.017	7.794	8.561	9.321	10.08	10.83	11.58	12.33	13.08	13.83	14.58
	u, Btu/lbm	—	1138.5	1177.2	1215.3	1253.6	1292.4	1332.0	1372.3	1413.5	1455.5	1498.4	1542.3	1587.0
	h, Btu/lbm	—	1230.6	1281.1	1330.7	1380.3	1430.4	1481.1	1532.6	1584.9	1638.1	1692.1	1747.0	1802.8
	s, Btu/(lbm·R)	—	1.6792	1.7348	1.7840	1.8287	1.8702	1.9089	1.9455	1.9801	2.0131	2.0447	2.0751	2.1043
100 (327.8)	v, ft³/lbm	—	4.934	5.587	6.216	6.834	7.445	8.053	8.657	9.260	9.861	10.46	11.06	11.66
	u, Btu/lbm	—	1136.2	1175.7	1214.2	1252.8	1291.8	1331.4	1371.9	1413.1	1455.2	1498.2	1542.0	1586.8
	h, Btu/lbm	—	1227.5	1279.1	1329.3	1379.2	1429.6	1480.5	1532.1	1584.5	1637.7	1691.8	1746.7	1802.5
	s, Btu/(lbm·R)	—	1.6519	1.7087	1.7584	1.8035	1.8451	1.8840	1.9206	1.9553	1.9884	2.0200	2.0504	2.0796
140 (353.1)	v, ft³/lbm	—	3.466	3.952	4.412	4.860	5.301	5.739	6.173	6.605	7.036	7.466	7.895	8.324
	u, Btu/lbm	—	1131.4	1172.7	1212.1	1251.2	1290.5	1330.4	1371.0	1412.4	1454.6	1497.7	1541.6	1586.4
	h, Btu/lbm	—	1221.2	1275.1	1326.4	1377.1	1427.9	1479.1	1531.0	1583.5	1636.9	1691.1	1746.1	1802.0
	s, Btu/(lbm·R)	—	1.6090	1.6684	1.7193	1.7650	1.8070	1.8461	1.8829	1.9178	1.9509	1.9826	2.0130	2.0423
180 (373.1)	v, ft³/lbm	—	2.648	3.042	3.409	3.763	4.110	4.453	4.793	5.131	5.467	5.802	6.137	6.471
	u, Btu/lbm	—	1126.2	1169.6	1210.0	1249.6	1289.3	1329.4	1370.2	1411.7	1454.0	1497.2	1541.2	1586.0
	h, Btu/lbm	—	1214.4	1270.9	1323.5	1374.9	1426.2	1477.7	1529.8	1582.6	1636.1	1690.4	1745.6	1801.5
	s, Btu/(lbm·R)	—	1.5751	1.6374	1.6895	1.7359	1.7783	1.8177	1.8546	1.8896	1.929	1.9546	1.9851	2.0144

(Continued)

Table C.3a Superheated Water Vapor (English Units) *continued*

p,psia (T_sat, °F)		400	500	600	700	800	900	1000	1100	1200	1300	1400	1500	1600
						Temperature, °F								
	v, ft³/lbm	2.361	2.724	3.058	3.379	3.693	4.003	4.310	4.615	4.918	5.220	5.521	5.822	6.123
200 (381.8)	u, Btu/lbm	1123.5	1168.0	1208.9	1248.8	1288.6	1328.9	1369.8	1411.4	1453.7	1496.9	1540.9	1585.8	1631.6
	h, Btu/lbm	1210.8	1268.8	1322.0	1373.8	1425.3	1477.0	1529.3	1582.1	1635.7	1690.1	1745.3	1801.3	1858.1
	s, Btu/(lbm·R)	1.5602	1.6240	1.6769	1.7236	1.7662	1.8057	1.8427	1.8778	1.9111	1.9429	1.9734	2.0027	2.0310
	v, ft³/lbm	–	2.150	2.426	2.688	2.943	3.193	3.440	3.685	3.929	4.172	4.414	4.655	4.896
250 (401.0)	u, Btu/lbm	–	1163.8	1206.1	1246.7	1287.0	1327.6	1368.7	1410.5	1453.0	1496.3	1540.4	1585.3	1631.1
	h, Btu/lbm	–	1263.3	1318.3	1371.1	1423.2	1475.3	1527.9	1581.0	1634.8	1689.3	1744.6	1800.7	1857.6
	s, Btu/(lbm·R)	–	1.5950	1.6496	1.6972	1.7403	1.7801	1.8174	1.8526	1.8860	1.9179	1.9485	1.9779	2.0062
	v, ft³/lbm	–	1.766	2.004	2.227	2.442	2.653	2.860	3.066	3.270	3.473	3.675	3.877	4.078
300 (417.4)	u, Btu/lbm	–	1159.5	1203.2	1244.6	1285.4	1326.3	1367.7	1409.6	1452.2	1495.6	1539.8	1584.8	1630.7
	h, Btu/lbm	–	1257.5	1314.5	1368.3	1421.0	1473.6	1526.4	1579.8	1633.8	1688.4	1743.8	1800.0	1857.0
	s, Btu/(lbm·R)	–	1.5703	1.6268	1.6753	1.7189	1.7591	1.7966	1.8319	1.8655	1.8975	1.9281	1.9575	1.9859
	v, ft³/lbm	–	1.284	1.476	1.650	1.816	1.978	2.136	2.292	2.446	2.599	2.752	2.904	3.055
400 (444.7)	u, Btu/lbm	–	1150.1	1197.3	1240.4	1282.1	1323.7	1365.5	1407.8	1450.7	1494.3	1538.7	1583.8	1629.8
	h, Btu/lbm	–	1245.2	1306.6	1362.5	1416.6	1470.1	1523.6	1577.4	1631.8	1686.8	1742.4	1798.8	1855.9
	s, Btu/(lbm·R)	–	1.5284	1.5894	1.6398	1.6846	1.7254	1.7634	1.7991	1.8329	1.8650	1.8958	1.9253	1.9537
	v, ft³/lbm	–	0.7947	0.9456	1.073	1.190	1.302	1.411	1.517	1.622	1.726	1.829	1.931	2.033
600 (486.3)	u, Btu/lbm	–	1128.0	1184.5	1231.5	1275.4	1318.4	1361.2	1404.2	1447.7	1491.7	1536.4	1581.8	1628.0
	h, Btu/lbm	–	1216.2	1289.5	1350.6	1407.6	1462.9	1517.8	1572.7	1627.8	1683.4	1739.5	1796.3	1853.7
	s, Btu/(lbm·R)	–	1.4594	1.5322	1.5874	1.6345	1.6768	1.7157	1.7521	1.7863	1.8188	1.8499	1.8796	1.9082
	v, ft³/lbm	–	–	0.6776	0.7829	0.8764	0.9640	1.048	1.130	1.210	1.289	1.367	1.445	1.522
800 (518.3)	u, Btu/lbm	–	–	1170.1	1222.1	1268.4	1312.9	1356.7	1400.5	1444.6	1489.1	1534.2	1579.8	1626.2
	h, Btu/lbm	–	–	1270.4	1338.0	1398.2	1455.6	1511.9	1567.8	1623.8	1680.0	1736.6	1793.7	1851.5
	s, Btu/(lbm·R)	–	–	1.4863	1.5473	1.5971	1.6410	1.6809	1.7180	1.7527	1.7856	1.8169	1.8469	1.8756
	v, ft³/lbm	–	–	0.5140	0.6080	0.6878	0.7610	0.8305	0.8976	0.9630	1.027	1.090	1.153	1.215
1000 (544.7)	u, Btu/lbm	–	–	1153.7	1212.0	1261.2	1307.3	1352.2	1396.8	1441.5	1486.4	1531.9	1577.8	1624.4
	h, Btu/lbm	–	–	1248.8	1324.5	1388.5	1448.1	1505.9	1562.9	1619.7	1676.5	1733.7	1791.2	1849.3
	s, Btu/(lbm·R)	–	–	1.4452	1.5137	1.5666	1.6122	1.6532	1.6910	1.7263	1.7595	1.7911	1.8212	1.8501

p.psia (T_sat °F)		800	900	1000	1100	1200	1300	1400	1500	1600	1700	1800	1900	2000
2000 (636.0)	v, ft³/lbm	0.3071	0.3534	0.3945	0.4325	0.4685	0.5031	0.5368	0.5697	0.6020	0.6340	0.6656	0.6971	0.7284
	u, Btu/lbm	1220.1	1276.8	1328.1	1377.2	1425.2	1472.7	1520.2	1567.6	1615.4	1663.5	1712.0	1761.0	1810.6
	h, Btu/lbm	1333.8	1407.6	1474.1	1537.2	1598.6	1659.0	1718.8	1778.5	1838.2	1898.1	1958.3	2019.0	2080.1
	s, Btu/(lbm·R)	1.4564	1.5128	1.5600	1.6019	1.6400	1.6753	1.7084	1.7397	1.7694	1.7978	1.8251	1.8513	1.8767
3000 (695.5)	v, ft³/lbm	0.1757	0.2160	0.2485	0.2772	0.3036	0.3285	0.3524	0.3754	0.3978	0.4198	0.4416	0.4631	0.4844
	u, Btu/lbm	1167.6	1241.8	1301.7	1356.2	1408.0	1458.5	1508.1	1557.3	1606.3	1655.3	1704.5	1754.0	1803.9
	h, Btu/lbm	1265.2	1361.7	1439.6	1510.0	1576.6	1640.8	1703.7	1765.6	1827.1	1888.4	1949.6	2011.1	2072.8
	s, Btu/(lbm·R)	1.3677	1.4416	1.4969	1.5436	1.5850	1.6226	1.6573	1.6897	1.7203	1.7494	1.7771	1.8037	1.8293
4000	v, ft³/lbm	0.1052	0.1462	0.1752	0.1995	0.2213	0.2414	0.2603	0.2784	0.2959	0.3129	0.3296	0.3462	0.3625
	u, Btu/lbm	1095.0	1201.5	1272.9	1333.9	1390.1	1443.7	1495.7	1546.7	1597.1	1647.2	1697.1	1747.1	1797.3
	h, Btu/lbm	1172.9	1309.7	1402.6	1481.6	1553.9	1622.4	1688.4	1752.8	1816.1	1878.8	1941.1	2003.3	2065.6
	s, Btu/(lbm·R)	1.2742	1.3791	1.4451	1.4975	1.5425	1.5825	1.6190	1.6528	1.6843	1.7140	1.7422	1.7691	1.7950
5000	v, ft³/lbm	0.05933	0.1038	0.1312	0.1530	0.1720	0.1892	0.2052	0.2203	0.2348	0.2489	0.2626	0.2761	0.2895
	u, Btu/lbm	987.2	1155.1	1242.0	1310.6	1371.6	1428.6	1483.2	1536.1	1587.9	1639.0	1689.7	1740.3	1790.8
	h, Btu/lbm	1042.1	1251.1	1363.4	1452.2	1530.8	1603.7	1673.0	1739.9	1805.2	1869.3	1932.7	1995.7	2058.6
	s, Btu/(lbm·R)	1.1586	1.3192	1.3990	1.4579	1.5068	1.5495	1.5878	1.6228	1.6553	1.6857	1.7144	1.7417	1.7678
6000	v, ft³/lbm	0.03942	0.07588	0.1021	0.1222	0.1393	0.1545	0.1685	0.1817	0.1942	0.2063	0.2180	0.2295	0.2409
	u, Btu/lbm	896.9	1102.9	1209.1	1286.4	1352.7	1413.3	1470.5	1525.4	1578.7	1630.9	1682.4	1733.4	1784.3
	h, Btu/lbm	940.6	1187.2	1322.4	1422.1	1507.3	1584.9	1657.6	1727.1	1794.3	1859.9	1924.5	1988.3	2051.7
	s, Btu/(lbm·R)	1.0710	1.2601	1.3563	1.4224	1.4754	1.5208	1.5610	1.5974	1.6309	1.6620	1.6912	1.7189	1.7452
7000	v, ft³/lbm	0.03341	0.05760	0.08172	0.1004	0.1161	0.1299	0.1425	0.1542	0.1653	0.1759	0.1862	0.1963	0.2062
	u, Btu/lbm	855.0	1049.7	1175.0	1261.7	1333.5	1397.8	1457.7	1514.6	1569.4	1622.8	1675.0	1726.7	1777.8
	h, Btu/lbm	898.3	1124.3	1280.9	1391.8	1483.9	1566.1	1642.3	1714.4	1783.5	1850.7	1916.3	1981.0	2045.0
	s, Btu/(lbm·R)	1.0321	1.2049	1.3163	1.3899	1.4471	1.4953	1.5374	1.5751	1.6096	1.6414	1.6711	1.6991	1.7257
8000	v, ft³/lbm	0.03061	0.04657	0.06722	0.08445	0.09892	0.1116	0.1231	0.1337	0.1437	0.1533	0.1625	0.1715	0.1803
	u, Btu/lbm	830.7	1003.7	1141.0	1236.8	1314.2	1382.3	1444.9	1503.8	1560.1	1614.6	1667.7	1719.9	1771.4
	h, Btu/lbm	876.0	1072.6	1240.5	1361.9	1460.6	1547.5	1627.1	1701.7	1772.9	1841.5	1908.3	1973.7	2038.4
	s, Btu/(lbm·R)	1.0098	1.1598	1.2793	1.3598	1.4212	1.4720	1.5160	1.5552	1.5906	1.6231	1.6533	1.6817	1.7085

Temperature, °F

Note: Saturated liquid entropies have been adjusted to make the Gibbs functions of the liquid and vapor phases exactly equal. For this reason, there are some small differences between values presented here and the original tables.

Sources: Reprinted by permission from Reynolds, W. C., Perkins, H. C., Engineering Thermodynamics, second ed., 1977, McGraw-Hill, New York. Recalculated from equations given in Keenan, J. H., Keyes, F. G., Hill, P. G., Moore, J. G. Steam Tables. Wiley, New York, 1969. Reprinted by permission of John Wiley & Sons, Inc.

Table C.3b Superheated Water Vapor (Metric Units)

p, MPa (T_sat °C)		50	100	150	200	250	300	350	400	500	600	700	800	900
								Temperature, °C						
0.002 (17.5)	v, m³/kg	74.52	86.08	97.63	109.2	120.7	132.3	143.8	155.3	178.4	201.5	224.6	247.6	270.7
	u, kJ/kg	2445.2	2516.3	2588.3	2661.6	2736.2	2812.2	2889.8	2969.0	3132.3	3302.5	3479.7	3663.9	3855.1
	h, kJ/kg	2594.3	2688.4	2783.6	2879.9	2977.6	3076.7	3177.4	3279.6	3489.1	3705.5	3928.8	4159.1	4396.5
	s, kJ/(kg·K)	8.9227	9.1936	9.4328	9.6479	9.8442	10.0251	10.1935	10.3513	10.6414	10.9044	11.1465	11.3718	11.5832
0.005 (32.9)	v, m³/kg	29.78	34.42	39.04	43.66	48.28	52.90	57.51	62.13	71.36	80.59	89.82	99.05	108.3
	u, kJ/kg	2444.7	2516.0	2588.1	2661.4	2736.1	2812.2	2889.8	2968.9	3132.3	3302.5	3479.6	3663.9	3855.0
	h, kJ/kg	2593.6	2688.1	2783.3	2879.8	2977.5	3076.6	3177.3	3279.6	3489.1	3705.4	3928.8	4159.1	4396.5
	s, kJ/(kg·K)	8.4982	8.7699	9.0095	9.2248	9.4212	9.6022	9.7706	9.9284	10.2185	10.4815	10.7236	10.9489	11.1603
0.01 (45.8)	v, m³/kg	14.87	17.20	19.51	21.83	24.14	26.45	28.75	31.06	35.68	40.29	44.91	49.53	54.14
	u, kJ/kg	2443.9	2515.5	2587.9	2661.3	2736.0	2812.1	2889.7	2968.8	3132.3	3302.5	3479.6	3663.8	3855.0
	h, kJ/kg	2592.6	2687.5	2783.0	2879.5	2977.3	3076.5	3177.2	3279.5	3489.0	3705.4	3928.7	4159.1	4396.4
	s, kJ/(kg·K)	8.1757	8.4487	8.6890	8.9046	9.1010	9.2821	9.4506	9.6084	9.8985	10.1616	10.4037	10.6290	10.8404
0.02 (60.1)	v, m³/kg	—	8.585	9.748	10.91	12.06	13.22	14.37	15.53	17.84	20.15	22.45	24.76	27.07
	u, kJ/kg	—	2514.5	2587.3	2660.9	2735.7	2811.9	2889.5	2968.5	3132.2	3302.4	3479.6	3663.8	3855.0
	h, kJ/kg	—	2686.9	2782.3	2879.1	2977.0	3076.3	3177.0	3279.4	3488.9	3705.3	3928.7	4159.1	4396.4
	s, kJ/(kg·K)	—	8.1263	8.3678	8.5839	8.7807	8.9619	9.1304	9.2884	9.5785	9.8417	10.0838	10.3091	10.5205
0.05 (81.3)	v, m³/kg	—	3.418	3.889	4.356	4.820	5.284	5.747	6.209	7.134	8.057	8.981	9.904	10.83
	u, kJ/kg	—	2511.6	2585.6	2659.8	2735.0	2811.3	2889.1	2968.4	3131.9	3302.2	3479.5	3663.7	3854.9
	h, kJ/kg	—	2682.5	2780.1	2877.6	2976.0	3075.5	3176.4	3278.9	3488.6	3705.1	3928.5	4158.9	4396.3
	s, kJ/(kg·K)	—	7.6955	7.9409	8.1588	8.3564	8.5380	8.7069	8.8650	9.1554	9.4186	9.6608	9.8861	10.0975
0.07 (89.9)	v, m³/kg	—	2.434	2.773	3.108	3.441	3.772	4.103	4.434	5.095	5.755	6.415	7.074	7.734
	u, kJ/kg	—	2509.6	2584.5	2659.1	2734.5	2811.0	2888.8	2968.2	3131.8	3302.0	3479.4	3663.6	3854.9
	h, kJ/kg	—	2680.0	2778.6	2876.7	2975.3	3075.0	3176.1	3278.6	3488.4	3704.9	3928.4	4158.8	4396.2
	s, kJ/(kg·K)	—	7.5349	7.7829	8.0020	8.2001	8.3821	8.5511	8.7094	8.9999	9.2632	9.5054	9.7307	9.9422
0.1 (99.6)	v, m³/kg	—	1.696	1.936	2.172	2.406	2.639	2.871	3.103	3.565	4.028	4.490	4.952	5.414
	u, kJ/kg	—	2506.6	2582.7	2658.0	2733.7	2810.4	2888.4	2967.8	3131.5	3301.9	3479.2	3663.5	3854.8
	h, kJ/kg	—	2676.2	2776.4	2875.3	2974.3	3074.3	3175.5	3278.1	3488.1	3704.7	3928.2	4158.7	4396.1
	s, kJ/(kg·K)	—	7.3622	7.6142	7.8351	8.0341	8.2165	8.3858	8.5442	8.8350	9.0984	9.3406	9.5660	9.7775

p, MPa (T_{sat}, °C)		150	200	250	300	350	400	450	500	550	600	700	800	900
						Temperature, °C								
0.15 (111.4)	v, m³/kg	1.285	1.444	1.601	1.757	1.912	2.067	2.222	2.376	2.530	2.685	2.993	3.301	3.609
	u, kJ/kg	2579.8	2656.2	2732.5	2809.5	2887.7	2967.3	3048.4	3131.1	3215.6	3301.6	3479.0	3663.4	3854.6
	h, kJ/kg	2772.6	2872.9	2972.7	3073.0	3174.5	3277.3	3381.7	3487.6	3595.1	3704.3	3927.9	4158.5	4395.9
	s, kJ/(kg·K)	7.4201	7.6441	7.8446	8.0278	8.1975	8.3562	8.5057	8.6473	8.7821	8.9109	9.1533	9.3787	9.5903
0.2 (120.2)	v, m³/kg	0.9596	1.080	1.199	1.316	1.433	1.549	1.665	1.781	1.897	2.013	2.244	2.475	2.706
	u, kJ/kg	2576.9	2654.4	2731.2	2808.6	2886.9	2966.7	3047.9	3130.7	3215.2	3301.4	3478.8	3663.2	3854.5
	h, kJ/kg	2768.8	2870.5	2971.0	3071.8	3173.5	3276.5	3381.0	3487.0	3594.7	3704.0	3927.7	4158.3	4395.8
	s, kJ/(kg·K)	7.2803	7.5074	7.7094	7.8934	8.0636	8.2226	8.3723	8.5140	8.6489	8.7778	9.0203	9.2458	9.4574
0.4 (143.6)	v, m³/kg	0.4708	0.5342	0.5951	0.6548	0.7139	0.7726	0.8311	0.8893	0.9475	1.006	1.121	1.237	1.353
	u, kJ/kg	2564.5	2646.8	2726.1	2804.8	2884.0	2964.4	3046.0	3129.2	3213.9	3300.2	3477.9	3662.5	3853.9
	h, kJ/kg	2752.8	2860.5	2964.2	3066.7	3169.6	3273.4	3378.4	3484.9	3592.9	3702.4	3926.5	4157.4	4395.1
	s, kJ/(kg·K)	6.9307	7.1714	7.3797	7.5670	7.7390	7.8992	8.0497	8.1921	8.3274	8.4566	8.6995	8.9253	9.1370
0.6 (158.9)	v, m³/kg	—	0.3520	0.3938	0.4344	0.4742	0.5137	0.5529	0.5920	0.6309	0.6697	0.7472	0.8245	0.9017
	u, kJ/kg	—	2638.9	2720.9	2801.0	2881.1	2962.0	3044.1	3127.6	3212.5	3299.1	3477.1	3661.8	3853.3
	h, kJ/kg	—	2850.1	2957.2	3061.6	3165.7	3270.2	3375.9	3482.7	3591.1	3700.9	3925.4	4156.5	4394.4
	s, kJ/(kg·K)	—	6.9673	7.1824	7.3732	7.5472	7.7086	7.8600	8.0029	8.1386	8.2682	8.5115	8.7375	8.9494
0.8 (170.4)	v, m³/kg	—	0.2608	0.2931	0.3241	0.3544	0.3843	0.4139	0.4433	0.4726	0.5018	0.5601	0.6181	0.6761
	u, kJ/kg	—	2630.6	2715.5	2797.1	2878.2	2959.7	3042.2	3125.9	3211.2	3297.9	3476.2	3661.1	3852.8
	h, kJ/kg	—	2839.2	2950.0	3056.4	3161.7	3267.1	3373.3	3480.6	3589.3	3699.4	3924.3	4155.7	4393.6
	s, kJ/(kg·K)	—	6.8167	7.0392	7.2336	7.4097	7.5723	7.7245	7.8680	8.0042	8.1341	8.3779	8.6041	8.8161
1 (179.9)	v, m³/kg	—	0.2060	0.2327	0.2579	0.2825	0.3066	0.3304	0.3541	0.3776	0.401 1	0.4478	0.4943	0.5407
	u, kJ/kg	—	2621.9	2709.9	2793.2	2875.2	2957.3	3040.2	3124.3	3209.8	3296.8	3475.4	3660.5	3852.2
	h, kJ/kg	—	2827.9	2942.6	3051.2	3157.7	3263.9	3370.7	3478.4	3587.5	3697.9	3923.1	4154.8	4392.9
	s, kJ/(kg·K)	—	6.6948	6.9255	7.1237	7.3019	7.4658	7.6188	7.7630	7.8996	8.0298	8.2740	8.5005	8.7127
1.5 (198.3)	v, m³/kg	—	0.1325	0.1520	0.1697	0.1866	0.2030	0.2192	0.2352	0.2510	0.2668	0.2981	0.3292	0.3603
	u, kJ/kg	—	2598.1	2695.3	2783.1	2867.6	2951.3	3035.3	3120.3	3206.4	3293.9	3473.2	3658.7	3850.8
	h, kJ/kg	—	2796.8	2923.2	3037.6	3147.4	3255.8	3364.1	3473.0	3582.9	3694.0	3920.3	4152.6	4391.2
	s, kJ/(kg·K)	—	6.4554	6.7098	6.9187	7.1025	7.2697	7.4249	7.5706	7.7083	7.8393	8.0846	8.3118	8.5243

(Continued)

Table C.3b Superheated Water Vapor (Metric Units)   *continued*

P, MPa (T_sat, °C)		250	300	350	400	450	500	550	600	650	700	750	800	900
	v, m³/kg	0.1114	0.1255	0.1386	0.1512	0.1635	0.1757	0.1877	0.1996	0.2114	0.2232	0.2350	0.2467	0.2700
2 (212.4)	u, kJ/kg	2679.6	2772.6	2859.8	2945.2	3030.4	3116.2	3203.0	3290.9	3380.2	3471.0	3563.2	3657.0	3849.3
	h, kJ/kg	2902.5	3023.5	3137.0	3247.6	3357.5	3467.6	3578.3	3690.1	3803.1	3917.5	4033.2	4150.4	4389.4
	s, kJ/(kg·K)	6.5461	6.7672	6.9571	7.1279	7.2853	7.4325	7.5713	7.7032	7.8290	7.9496	8.0656	8.1774	8.3903
	v, m³/kg	0.07058	0.08114	0.09053	0.09936	0.1079	0.1162	0.1244	0.1324	0.1404	0.1484	0.1563	0.1641	0.1798
3 (233.9)	u, kJ/kg	2644.0	2750.0	2843.7	2932.7	3020.4	3107.9	3196.0	3285.0	3375.2	3466.6	3559.4	3653.6	3846.5
	h, kJ/kg	2855.8	2993.5	3115.3	3230.8	3344.0	3456.5	3569.1	3682.3	3796.5	3911.7	4028.2	4146.0	4385.9
	s, kJ/(kg·K)	6.2880	6.5398	6.7436	6.9220	7.0842	7.2346	7.3757	7.5093	7.6364	7.7580	7.8747	7.9971	8.2008
	v, m³/kg	–	0.05884	0.06645	0.07341	0.08003	0.08643	0.09269	0.09885	0.1049	0.1109	0.1169	0.1229	0.1347
4 (250.4)	u, kJ/kg	–	2725.3	2826.6	2919.9	3010.1	3099.5	3189.0	3279.1	3370.1	3462.1	3555.5	3650.1	3843.6
	h, kJ/kg	–	2960.7	3092.4	3213.5	3330.2	3445.2	3559.7	3674.4	3789.8	3905.9	4023.2	4141.6	4382.3
	s, kJ/(kg·K)	–	6.3622	6.5828	6.7698	6.9371	7.0908	7.2343	7.3696	7.4981	7.6206	7.7381	7.8511	8.0655
	v, m³/kg	–	0.03616	0.04223	0.04739	0.05214	0.05665	0.06101	0.06525	0.06942	0.07352	0.07758	0.08160	0.08958
6 (275.6)	u, kJ/kg	–	2667.2	2789.6	2892.8	2986.9	3082.2	3174.6	3266.9	3359.6	3453.2	3547.6	3643.1	3837.8
	h, kJ/kg	–	2884.2	3043.0	3177.2	3301.8	3422.1	3540.6	3658.4	3776.2	3894.3	4013.1	4132.7	4375.3
	s, kJ/(kg·K)	–	6.0682	6.3342	6.5415	6.7201	6.8811	7.0296	7.1685	7.2996	7.4242	7.5433	7.6675	7.8735
	v, m³/kg	–	0.02426	0.02995	0.03432	0.03817	0.04175	0.04516	0.04845	0.05166	0.05481	0.05791	0.06097	0.06702
8 (295.1)	u, kJ/kg	–	2590.9	2747.7	2863.8	2966.7	3064.3	3159.8	3254.4	3349.0	3444.0	3539.6	3636.1	3832.1
	h, kJ/kg	–	2785.0	2987.3	3138.3	3272.0	3398.3	3521.0	3642.0	3762.3	3882.5	4002.9	4123.8	4368.3
	s, kJ/(kg·K)	–	5.7914	6.1309	6.3642	6.5559	6.7248	6.8786	7.0214	7.1553	7.2821	7.4027	7.5182	7.7359
	v, m³/kg	–	–	0.02242	0.02641	0.02975	0.03279	0.03564	0.03837	0.04101	0.04358	0.04611	0.04859	0.05349
10 (311.1)	u, kJ/kg	–	–	2699.2	2832.4	2943.3	3045.8	3144.5	3241.7	3338.2	3434.7	3531.5	3629.0	3826.3
	h, kJ/kg	–	–	2923.4	3096.5	3240.8	3373.6	3500.9	3625.3	3748.3	3870.5	3992.6	4114.9	4361.2
	s, kJ/(kg·K)	–	–	5.9451	6.2127	6.4197	6.5974	6.7569	6.9037	7.0406	7.1696	7.2919	7.4086	7.6280
	v, m³/kg	–	–	0.01721	0.02108	0.02412	0.02680	0.02929	0.03164	0.03390	0.03610	0.03824	0.04034	0.04447
12 (324.8)	u, kJ/kg	–	–	2641.1	2798.3	2918.8	3026.6	3128.9	3228.7	3327.2	3425.3	3523.4	3621.8	3820.6
	h, kJ/kg	–	–	2847.6	3051.2	3208.2	3348.2	3480.3	3608.3	3734.0	3858.4	3982.3	4105.9	4354.2
	s, kJ/(kg·K)	–	–	5.7604	6.0754	6.3006	6.4879	6.6535	6.8045	6.9445	7.0757	7.1998	7.3178	7.5390

Temperature, °C

P, MPa (T_sat °C)		Temperature, °C													
		400	450	500	550	600	650	700	750	800	850	900	950	1000
15 (342.2)	v, m³/kg	0.01565	0.01845	0.02080	0.02293	0.02491	0.02680	0.02861	0.03037	0.03210	0.03379	0.03546	0.03711	0.03875
	u, kJ/kg	2740.7	2879.5	2996.5	3104.7	3208.6	3310.4	3410.9	3511.0	3611.0	3711.2	3811.9	3913.2	4015.4
	h, kJ/kg	2975.4	3156.2	3308.5	3448.6	3582.3	3712.3	3840.1	3966.6	4092.4	4218.0	4343.8	4469.9	4596.6
	s, kJ/(kg·K)	5.8819	6.1412	6.3451	6.5207	6.6784	6.8232	6.9580	7.0848	7.2048	7.3192	7.4288	7.5340	7.6356
20 (365.8)	v, m³/kg	0.00994	0.01270	0.01477	0.01656	0.01818	0.01969	0.02113	0.02251	0.02385	0.02516	0.02645	0.02771	0.02897
	u, kJ/kg	2619.2	2806.2	2942.8	3062.3	3174.0	3281.5	3386.5	3490.0	3592.7	3695.1	3797.4	3900.0	4003.1
	h, kJ/kg	2818.1	3060.1	3238.2	3393.4	3537.6	3675.3	3809.1	3940.3	4069.8	4198.3	4326.4	4454.3	4582.5
	s, kJ/(kg·K)	5.5548	5.9025	6.1409	6.3356	6.5056	6.6591	6.8002	6.9317	7.0553	7.1723	7.2839	7.3907	7.4933
22.088 (374.136)	v, m³/kg	0.00818	0.01104	0.01305	0.01475	0.01627	0.01768	0.01901	0.02029	0.02152	0.02272	0.02389	0.02505	0.02619
	u, kJ/kg	2552.9	2772.1	2919.0	3043.9	3159.1	3269.1	3376.1	3481.1	3585.0	3688.3	3791.4	3894.5	3998.0
	h, kJ/kg	2733.7	3015.9	3207.2	3369.6	3518.4	3659.6	3796.0	3929.2	4060.3	4190.1	4319.1	4447.9	4576.6
	s, kJ/(kg·K)	5.4013	5.8072	6.0634	6.2670	6.4426	6.5998	6.7437	6.8772	7.0024	7.1206	7.2330	7.3404	7.4436
30	v, m³/kg	0.00279	0.00674	0.00868	0.01017	0.01145	0.01260	0.01366	0.01466	0.01562	0.01655	0.01745	0.01833	0.01920
	u, kJ/kg	2067.3	2619.3	2820.7	2970.3	3100.5	3221.0	3335.8	3447.0	3555.6	3662.6	3768.5	3873.8	3978.8
	h, kJ/kg	2151.0	2821.4	3081.0	3275.4	3443.9	3598.9	3745.7	3886.9	4024.3	4159.0	4291.9	4423.6	4554.7
	s, kJ/(kg·K)	4.4736	5.4432	5.7912	6.0350	6.2339	6.4066	6.5614	6.7030	6.8341	6.9568	7.0726	7.1825	7.2875
40	v, m³/kg	0.00191	0.00369	0.00562	0.00698	0.00809	0.00906	0.00994	0.01076	0.01152	0.01226	0.01296	0.01365	0.01432
	u, kJ/kg	1854.5	2365.1	2678.4	2869.7	3022.6	3158.0	3283.6	3402.9	3517.9	3629.8	3739.4	3847.5	3954.6
	h, kJ/kg	1930.8	2512.8	2903.3	3149.1	3346.4	3520.6	3681.3	3833.1	3978.8	4120.0	4257.9	4393.6	4527.6
	s, kJ/(kg·K)	4.1143	4.9467	5.4707	5.7793	6.0122	6.2063	6.3759	6.5281	6.6671	6.7957	6.9158	7.0291	7.1365
60	v, m³/kg	0.00163	0.00208	0.00296	0.00396	0.00483	0.00560	0.00627	0.00689	0.00746	0.00800	0.00851	0.00900	0.00948
	u, kJ/kg	1745.3	2053.9	2390.5	2658.8	2861.1	3028.8	3177.2	3313.6	3441.6	3563.6	3681.0	3795.0	3906.4
	h, kJ/kg	1843.4	2179.0	2567.9	2896.2	3151.2	3364.5	3553.6	3726.8	3889.1	4043.3	4191.5	4335.0	4475.2
	s, kJ/(kg·K)	3.9325	4.4128	4.9329	5.3449	5.6460	5.8838	6.0832	6.2569	6.4118	6.5523	6.6814	6.8012	6.9135
80	v, m³/kg	0.00152	0.00177	0.00219	0.00276	0.00339	0.00398	0.00452	0.00502	0.00548	0.00591	0.00632	0.00671	0.00709
	u, kJ/kg	1687.0	1944.9	2218.9	2483.9	2711.8	2904.7	3073.2	3225.3	3365.7	3497.3	3622.3	3742.1	3857.8
	h, kJ/kg	1808.3	2086.9	2393.9	2704.9	2982.7	3222.8	3434.7	3626.6	3803.8	3970.1	4127.9	4279.1	4425.2
	s, kJ/(kg·K)	-3.8338	4.2328	4.6432	5.0331	5.3609	5.6284	5.8521	6.0445	6.2137	6.3652	6.5026	6.6289	6.7459

Note: Saturated liquid entropies have been adjusted to make the Gibbs functions of the liquid and vapor phases exactly equal. For this reason, there are some small differences between values presented here and the original tables.
Sources: Reprinted by permission from Reynolds, W. C., Perkins, H. C. Engineering Thermodynamics, second ed., 1977, McGraw-Hill, New York. Recalculated from equations given in Keenan, J. H., Keyes, F. G., Hill, P. G., Moore, J. G. Steam Tables. Wiley, New York, 1969. Reprinted by permission of John Wiley & Sons, Inc.

Table C.4a Compressed Liquid Water (English Units)

Temp. °F	p = 500 psia (467.13°F)				p = 1000 psia (544.75°F)				p = 1500 psia (596.39°F)			
	v ft³/lbm	u Btu/lbm	h Btu/lbm	s Btu/(lbm·R)	v ft³/lbm	u Btu/lbm	h Btu/lbm	s Btu/(lbm·R)	v ft³/lbm	u Btu/lbm	h Btu/lbm	s Btu/(lbm·R)
Sat	0.019 748	447.70	449.53	0.649 04	0.021 591	538.39	542.38	0.743 20	0.023 461	604.97	611.48	0.808 24
32	0.015 994	0.00	1.49	0.000 00	0.015 967	0.03	2.99	0.000 05	0.015 939	0.05	4.47	0.000 07
50	0.015 998	18.02	19.50	0.035 99	0.015 972	17.99	20.94	0.035 92	0.015 946	17.95	22.38	0.035 84
100	0.016 106	67.87	69.36	0.129 32	0.016 082	67.70	70.68	0.129 01	0.016 058	67.53	71.99	0.128 70
150	0.016 318	117.66	119.17	0.214 57	0.016 293	117.38	120.40	0.214 10	0.016 268	117.10	121.62	0.213 64
200	0.016 608	167.65	169.19	0.293 41	0.016 580	167.26	170.32	0.292 81	0.016 554	166.87	171.46	0.292 21
250	0.016 972	217.99	219.56	0.367 02	0.016 941	217.47	220.61	0.366 28	0.016 910	216.96	221.65	0.365 54
300	0.017 416	268.92	270.53	0.436 41	0.017 379	268.24	271.46	0.435 52	0.017 343	267.58	272.39	0.434 63
350	0.017 954	320.71	322.37	0.502 49	0.017 909	319.83	323.15	0.501 40	0.017 865	318.98	323.94	0.500 34
400	0.018 608	373.68	375.40	0.566 04	0.018 550	372.55	375.98	0.564 72	0.018 493	371.45	376.59	0.563 43
450	0.019 420	428.40	430.19	0.627 98	0.019 340	426.89	430.47	0.626 32	0.019 264	425.44	430.79	0.624 70
500	—	—	—	—	0.020 36	483.8	487.5	0.6874	0.020 24	481.8	487.4	0.6853
550	—	—	—	—	—	—	—	—	0.021 58	542.1	548.1	0.7469

Temp. °F	p = 2000 psia (636.00°F)				p = 3000 psia (695.52°F)				p = 5000 psia			
	v ft³/lbm	u Btu/lbm	h Btu/lbm	s Btu/(lbm·R)	v ft³/lbm	u Btu/lbm	h Btu/lbm	s Btu/(lbm·R)	v ft³/lbm	u Btu/lbm	h Btu/lbm	s Btu/(lbm·R)
Sat	0.025 649	662.40	671.89	0.862 27	0.034 310	783.45	802.50	0.973 20				
32	0.015 912	0.06	5.95	0.000 08	0.015 859	0.09	8.90	0.000 09	0.015 755	0.11	14.70	−0.000 01
50	0.015 920	17.91	23.81	0.035 75	0.015 870	17.84	26.65	0.035 55	0.015 773	17.67	32.26	0.035 08
100	0.016 034	67.37	73.30	0.12839	0.015 987	67.04	75.91	0.127 77	0.015 897	66.40	81.11	0.126 51
200	0.016 527	166.49	172.60	0.291 62	0.016 476	165.74	174.89	0.290 46	0.016 376	164.32	179.47	0.288 18
300	0.017 308	266.93	273.33	0.433 76	0.017 240	265.66	275.23	0.432 05	0.017 110	263.25	279.08	0.428 75
400	0.018 439	370.38	377.21	0.562 16	0.018 334	368.32	378.50	0.559 70	0.018 141	364.47	381.25	0.555 06
450	0.019 191	424.04	431.14	0.623 13	0.019 053	421.36	431.93	0.620 11	0.018 803	416.44	433.84	0.614 51
500	0.020 14	479.8	487.3	0.6832	0.019 944	476.2	487.3	0.6794	0.019 603	469.8	487.9	0.6724
560	0.021 72	551.8	559.8	0.7565	0.021 382	546.2	558.0	0.7508	0.020 835	536.7	556.0	0.7411
600	0.023 30	605.4	614.0	0.8086	0.022 74	597.0	609.6	0.8004	0.021 91	584.0	604.2	0.7876
640	—	—	—	—	0.024 75	654.3	668.0	0.8545	0.023 34	634.6	656.2	0.8357
680	—	—	—	—	0.028 79	728.4	744.3	0.9226	0.025 35	690.6	714.1	0.8873
700	—	—	—	—	—	—	—	—	0.026 76	721.8	746.6	0.9156

Source: Reprinted from Van Wylen, G. J., Sonntag, R. E., 1986. Fundamentals of Classical Thermodynamics, third ed. Wiley, New York. Reprinted by permission of John Wiley & Sons, Inc.

Table C.4b Compressed Liquid Water (Metric Units)

Temp. °C	v m³/kg	u kJ/kg	h kJ/kg	s kJ/(kg·K)	v m³/kg	u kJ/kg	h kJ/kg	s kJ/(kg·K)	v m³/kg	u kJ/kg	h kJ/kg	s kJ/(kg·K)
	p = 5 MPa (263.99°C)				p = 10 MPa (311.06°C)				p = 15 MPa (342.24°C)			
Sat.	.001 285 9	1147.8	1154.2	2.9202	.001 452 4	1393.0	1407.6	3.3596	.001 658 1	1585.6	1610.5	3.6848
0	.000 997 7	.04	5.04	.0001	.000 995 2	.09	10.04	.0002	.000 992 8	.15	15.05	.0004
20	.000 999 5	83.65	88.65	.2956	.000 997 2	83.36	93.33	.2945	.000 995 0	83.06	97.99	.2934
40	.001 005 6	166.95	171.97	.5705	.001 003 4	166.35	176.38	.5686	.001 001 3	165.76	180.78	.5666
60	.001 014 9	250.23	255.30	.8285	.001 012 7	249.36	259.49	.8258	.001 010 5	248.51	263.67	.8232
80	.001 026 8	333.72	338.85	1.0720	.001 024 5	332.59	342.83	1.0688	.001 022 2	331.48	346.81	1.0656
100	.001 041 0	417.52	422.72	1.3030	.001 038 5	416.12	426.50	1.2992	.001 036 1	414.74	430.28	1.2955
120	.001 057 6	501.80	507.09	1.5233	.001 054 9	500.08	510.64	1.5189	.001 052 2	498.40	514.19	1.5145
140	.001 076 8	586.76	592.15	1.7343	.001 073 7	584.68	595.42	1.7292	.001 070 7	582.66	598.72	1.7242
160	.001 098 8	672.62	678.12	1.9375	.001 095 3	670.13	681.08	1.9317	.001 091 8	667.71	684.09	1.9260
180	.001 124 0	759.63	765.25	2.1341	.001 119 9	756.65	767.84	2.1275	.001 115 9	753.76	770.50	2.1210
200	.001 153 0	848.1	853.9	2.3255	.001 148 0	844.5	856.0	2.3178	.001 143 3	841.0	858.2	2.3104
220	.001 186 6	938.4	944.4	2.5128	.001 180 5	934.1	945.9	2.5039	.001 174 8	929.9	947.5	2.4953
240	.001 226 4	1031.4	1037.5	2.6979	.001 218 7	1026.0	1038.1	2.6872	.001 211 4	1020.8	1039.0	2.6771
260	.001 274 9	1127.9	1134.3	2.8830	.001 264 5	1121.1	1133.7	2.8699	.001 255 0	1114.6	1133.4	2.8576
280	–	–	–	–	.001 321 6	1220.9	1234.1	3.0548	.001 308 4	1212.5	1232.1	3.0393
300	–	–	–	–	.001 397 2	1328.4	1342.3	3.2469	.001 377 0	1316.6	1337.3	3.2260
320	–	–	–	–	–	–	–	–	.001 472 4	1431.1	1453.2	3.4247
340	–	–	–	–	–	–	–	–	.001 631 1	1567.5	1591.9	3.6546

(Continued)

Table C.4b Compressed Liquid Water (Metric Units) *continued*

Temp. °C	p = 20 MPa (365.81°C)				p = 30 MPa				p = 50 MPa			
	v m³/kg	u kJ/kg	h kJ/kg	s kJ/(kg·K)	v m³/kg	u kJ/kg	h kJ/kg	s kJ/(kg·K)	v m³/kg	u kJ/kg	h kJ/kg	s kJ/(kg·K)
Sat.	.002 036	1785.6	1826.3	4.0139	—	—	—	—	—	—	—	—
0	.000 990 4	.19	20.01	.0004	.000 985 6	.25	29.82	.0001	.000 976 6	.20	49.03	-.0014
20	.000 992 8	82.77	102.62	.2923	.000 988 6	82.17	111.84	.2899	.000 980 4	81.00	130.02	.2848
40	.000 999 2	165.17	185.16	.5646	.000 995 1	164.04	193.89	.5607	.000 987 2	161.86	211.21	.5527
60	.001 008 4	247.68	267.85	.8206	.001 004 2	246.06	276.19	.8154	.000 996 2	242.98	292.79	.8052
80	.001 019 9	330.40	350.80	1.0624	.001 015 6	328.30	358.77	1.0561	.001 007 3	324.34	374.70	1.0440
100	.001 033 7	413.39	434.06	1.2917	.001 029 0	410.78	441.66	1.2844	.001 020 1	405.88	456.89	1.2703
120	.001 049 6	496.76	517.76	1.5102	.001 044 5	493.59	524.93	1.5018	.001 034 8	487.65	539.39	1.4857
140	.001 067 8	580.69	602.04	1.7193	.001 062 1	576.88	608.75	1.7098	.001 051 5	569.77	622.35	1.6915
160	.001 088 5	665.35	687.12	1.9204	.001 082 1	660.82	693.28	1.9096	.001 070 3	652.41	705.92	1.8891
180	.001 112 0	750.95	773.20	2.1147	.001 104 7	745.59	778.73	2.1024	.001 091 2	735.69	790.25	2.0794
200	.001 138 8	837.7	860.5	2.3031	.001 130 2	831.4	865.3	2.2893	.001 114 6	819.7	875.5	2.2634
220	.001 169 3	925.9	949.3	2.4870	.001 159 0	918.3	953.1	2.4711	.001 140 8	904.7	961.7	2.4419
240	.001 204 6	1016.0	1040.0	2.6674	.001 192 0	1006.9	1042.6	2.6490	.001 170 2	990.7	1049.2	2.6158
260	.001 246 2	1108.6	1133.5	2.8459	.001 230 3	1097.4	1134.3	2.8243	.001 203 4	1078.1	1138.2	2.7860
280	.001 296 5	1204.7	1230.6	3.0248	.001 275 5	1190.7	1229.0	2.9986	.001 241 5	1167.2	1229.3	2.9537
300	.001 359 6	1306.1	1333.3	3.2071	.001 330 4	1287.9	1327.8	3.1741	.001 286 0	1258.7	1323.0	3.1200
320	.001 443 7	1415.7	1444.6	3.3979	.001 399 7	1390.7	1432.7	3.3539	.001 338 8	1353.3	1420.2	3.2868
340	.001 568 4	1539.7	1571.0	3.6075	.001 492 0	1501.7	1546.5	3.5426	.001 403 2	1452.0	1522.1	3.4557
360	.001 822 6	1702.8	1739.3	3.8772	.001 626 5	1626.6	1675.4	3.7494	.001 483 8	1556.0	1630.2	3.6291
380	—				.001 869 1	1781.4	1837.5	4.0012	.001 588 4	1667.2	1746.6	3.8101

Source: Reprinted from Van Wylen, G. J., Sonntag, R. E., 1986. Fundamentals of Classical Thermodynamics, third ed. Wiley, New York. Reprinted by permission of John Wiley & Sons, Inc.

Table C.5a Saturated Ammonia (English Units)

Temp. °F	Abs. Press. psia p	Specific Volume ft³/lbm			Enthalpy Btu/lbm			Entropy Btu/(lbm·R)		
		Sat. Liquid v_f	Evap. v_{fg}	Sat. Vapor v_g	Sat. Liquid h_f	Evap. h_{fg}	Sat. Vapor h_g	Sat. Liquid s_f	Evap. s_{fg}	Sat. Vapor s_g
-60	5.55	0.022 80	44.707	44.73	-21.2	610.8	589.6	-0.0517	1.5286	1.4769
-55	6.54	0.022 90	38.357	38.38	-15.9	607.5	591.6	-0.0386	1.5017	1.4631
-50	7.67	0.023 00	33.057	33.08	-10.6	604.3	593.7	-0.0256	1.4753	1.4497
-45	8.95	0.023 10	28.597	28.62	-5.3	600.9	595.6	-0.0127	1.4495	1.4368
-40	10.41	0.023 22	24.837	24.86	0	597.6	597.6	0.000	1.4242	1.4242
-35	12.05	0.023 33	21.657	21.68	5.3	594.2	599.5	0.0126	1.3994	1.4120
-30	13.90	0.023 50	18.947	18.97	10.7	590.7	601.4	0.0250	1.3751	1.4001
-25	15.98	0.023 60	16.636	16.66	16.0	587.2	603.2	0.0374	1.3512	1.3886
-20	18.30	0.023 70	14.656	14.68	21.4	583.6	605.0	0.0497	1.3277	1.3774
-15	20.88	0.023 81	12.946	12.97	26.7	580.0	606.7	0.0618	1.3044	1.3664
-10	23.74	0.023 93	11.476	11.50	32.1	576.4	608.5	0.0738	1.2820	1.3558
-5	26.92	0.024 06	10.206	10.23	37.5	572.6	610.1	0.0857	1.2597	1.3454
0	30.42	0.024 19	9.092	9.116	42.9	568.9	611.8	0.0975	1.2377	1.3352
5	34.27	0.024 32	8.1257	8.150	48.3	565.0	613.3	0.1092	1.2161	1.3253
10	38.51	0.024 46	7.2795	7.304	53.8	561.1	614.9	0.1208	1.1949	1.3157
15	43.14	0.024 60	6.5374	6.562	59.2	557.1	616.3	0.1323	1.1739	1.3062
20	48.21	0.024 74	5.8853	5.910	64.7	553.1	617.8	0.1437	1.1532	1.2969
25	53.73	0.024 88	5.3091	5.334	70.2	548.9	619.1	0.1551	1.1328	1.2879
30	59.74	0.025 03	4.8000	4.825	75.7	544.8	620.5	0.1663	1.1127	1.2790
35	66.26	0.025 18	4.3478	4.373	81.2	540.5	621.7	0.1775	1.0929	1.2704
40	73.32	0.025 33	3.9457	3.971	86.8	536.2	623.0	0.1885	1.0733	1.2618
45	80.96	0.025 48	3.5885	3.614	92.3	531.8	624.1	0.1996	1.0539	1.2535
50	89.19	0.025 64	3.2684	3.294	97.9	527.3	625.2	0.2105	1.0348	1.2453
55	98.06	0.025 81	2.9822	3.008	103.5	522.8	626.3	0.2214	1.0159	1.2373
60	107.6	0.025 97	2.7250	2.751	109.2	518.1	627.3	0.2322	0.9972	1.2294
65	117.8	0.026 14	2.4939	2.520	114.8	513.4	628.2	0.2430	0.9786	1.2216
70	128.8	0.026 32	2.2857	2.312	120.5	508.6	629.1	0.2537	0.9603	1.2140
75	140.5	0.026 50	2.0985	2.125	126.2	503.7	629.9	0.2643	0.9422	1.2065
80	153.0	0.026 68	1.9283	1.955	132.0	498.7	630.7	0.2749	0.9242	1.1991
85	166.4	0.026 87	1.7741	1.801	137.8	493.6	631.4	0.2854	0.9064	1.1918
90	180.6	0.027 07	1.6339	1.661	143.5	488.5	632.0	0.2958	0.8888	1.1846
95	195.8	0.027 27	1.5067	1.534	149.4	483.2	632.6	0.3062	0.8713	1.1775
100	211.9	0.027 47	1.3915	1.419	155.2	477.8	633.0	0.3166	0.8539	1.1705
105	228.9	0.027 69	1.2853	1.313	161.1	472.3	633.4	0.3269	0.8366	1.1635
110	247.0	0.027 90	1.1891	1.217	167.0	466.7	633.7	0.3372	0.8194	1.1566
115	266.2	0.028 13	1.0999	1.128	173.0	460.9	633.9	0.3474	0.8023	1.1497
120	286.4	0.028 36	1.0186	1.047	179.0	455.0	634.0	0.3576	0.7851	1.1427
125	307.8	0.028 60	0.9444	0.973	185.1	448.9	634.0	0.3679	0.7679	1.1358

Source: Reprinted from Van Wylen, G. J., Sonntag, R. E., 1986. Fundamentals of Classical Thermodynamics, third ed. Wiley, New York. Reprinted by permission of John Wiley & Sons, Inc.

Table C.5b Saturated Ammonia (Metric Units)

Temp. °C	Abs. Press. kPa p	Specific volume m³/kg			Enthalpy kJ/kg			Entropy kJ/(kg·K)		
		Sat. Liquid v_f	Evap. v_{fg}	Sat. vapor v_g	Sat. Liquid h_f	Evap. h_{fg}	Sat. Vapor h_g	Sat. Liquid s_f	Evap. s_{fg}	Sat. Vapor s_g
−50	40.88	0.001 424	2.6239	2.6254	−44.3	1416.7	1372.4	−0.1942	6.3502	6.1561
−48	45.96	0.001 429	2.3518	2.3533	−35.5	1411.3	1375.8	−0.1547	6.2696	6.1149
−46	51.55	0.001 434	2.1126	2.1140	−26.6	1405.8	1379.2	−0.1156	6.1902	6.0746
−44	57.69	0.001 439	1.9018	1.9032	−17.8	1400.3	1382.5	−0.0768	6.1120	6.0352
−42	64.42	0.001 444	1.7155	1.7170	−8.9	1394.7	1385.8	−0.0382	6.0349	5.9967
−40	71.77	0.001 449	1.5506	1.5521	0.0	1389.0	1389.0	0.0000	5.9589	5.9589
−38	79.80	0.001 454	1.4043	1.4058	8.9	1383.3	1392.2	0.0380	5.8840	5.9220
−36	88.54	0.001 460	1.2742	1.2757	17.8	1377.6	1395.4	0.0757	5.8101	5.8858
−34	98.05	0.001 465	1.1582	1.1597	26.8	1371.8	1398.5	0.1132	5.7372	5.8504
−32	108.37	0.001 470	1.0547	1.0562	35.7	1365.9	1401.6	0.1504	5.6652	5.8156
−30	119.55	0.001 476	0.9621	0.9635	44.7	1360.0	1404.6	0.1873	5.5942	5.7815
−28	131.64	0.001 481	0.8790	0.8805	53.6	1354.0	1407.6	0.2240	5.5241	5.7481
−26	144.70	0.001 487	0.8044	0.8059	62.6	1347.9	1410.5	0.2605	5.4548	5.7153
−24	158.78	0.001 492	0.7373	0.7388	71.6	1341.8	1413.4	0.2967	5.3864	5.6831
−22	173.93	0.001 498	0.6768	0.6783	80.7	1335.6	1416.2	0.3327	5.3188	5.6515
−20	190.22	0.001 504	0.6222	0.6237	89.7	1329.3	1419.0	0.3684	5.2520	5.6205
−18	207.71	0.001 510	0.5728	0.5743	98.8	1322.9	1421.7	0.4040	5.1860	5.5900
−16	226.45	0.001 515	0.5280	0.5296	107.8	1316.5	1424.4	0.4393	5.1207	5.5600
−14	246.51	0.001 521	0.4874	0.4889	116.9	1310.0	1427.0	0.4744	5.0561	5.5305
−12	267.95	0.001 528	0.4505	0.4520	126.0	1303.5	1429.5	0.5093	4.9922	5.5015
−10	290.85	0.001 534	0.4169	0.4185	135.2	1296.8	1432.0	0.5440	4.9290	5.4730
−8	315.25	0.001 540	0.3863	0.3878	144.3	1290.1	1434.4	0.5785	4.8664	5.4449
−6	341.25	0.001 546	0.3583	0.3599	153.5	1283.3	1436.8	0.6128	4.8045	5.4173
−4	368.90	0.001 553	0.3328	0.3343	162.7	1276.4	1439.1	0.6469	4.7432	5.3901
−2	398.27	0.001 559	0.3094	0.3109	171.9	1269.4	1441.3	0.6808	4.6825	5.3633
0	429.44	0.001 566	0.2879	0.2895	181.1	1262.4	1443.5	0.7145	4.6223	5.3369
2	462.49	0.001 573	0.2683	0.2698	190.4	1255.2	1445.6	0.7481	4.5627	5.3108

Temp. °C	Abs. Press. kPa p	Specific volume m³/kg			Enthalpy kJ/kg			Entropy kJ/(kg·K)		
		Sat. Liquid v_f	Evap. v_{fg}	Sat. vapor v_g	Sat. Liquid h_f	Evap. h_{fg}	Sat. Vapor h_g	Sat. Liquid s_f	Evap. s_{fg}	Sat. Vapor s_g
4	497.49	0.001 580	0.2502	0.2517	199.6	1248.0	1447.6	0.7815	4.5037	5.2852
6	534.51	0.001 587	0.2335	0.2351	208.9	1240.6	1449.6	0.8148	4.4451	5.2599
8	573.64	0.001 594	0.2182	0.2198	218.3	1233.2	1451.5	0.8479	4.3871	5.2350
10	614.95	0.001 601	0.2040	0.2056	227.6	1225.7	1453.3	0.8808	4.3295	5.2104
12	658.52	0.001 608	0.1910	0.1926	237.0	1218.1	1455.1	0.9136	4.2725	5.1861
14	704.44	0.001 616	0.1789	0.1805	246.4	1210.4	1456.8	0.9463	4.2159	5.1621
16	752.79	0.001 623	0.1677	0.1693	255.9	1202.6	1458.5	0.9788	4.1597	5.1385
18	803.66	0.001 631	0.1574	0.1590	265.4	1194.7	1460.0	1.0112	4.1039	5.1151
20	857.12	0.001 639	0.1477	0.1494	274.9	1186.7	1461.5	1.0434	4.0486	5.0920
22	913.27	0.001 647	0.1388	0.1405	284.4	1178.5	1462.9	1.0755	3.9937	5.0692
24	972.19	0.001 655	0.1305	0.1322	294.0	1170.3	1464.3	1.1075	3.9392	5.0467
26	1033.97	0.001 663	0.1228	0.1245	303.6	1162.0	1465.6	1.1394	3.8850	5.0244
28	1098.71	0.001 671	0.1156	0.1173	313.2	1153.6	1466.8	1.1711	3.8312	5.0023
30	1166.49	0.001 680	0.1089	0.1106	322.9	1145.0	1467.9	1.2028	3.7777	4.9805
32	1237.41	0.001 689	0.1027	0.1044	332.6	1136.4	1469.0	1.2343	3.7246	4.9589
34	1311.55	0.001 698	0.0969	0.0986	342.3	1127.6	1469.9	1.2656	3.6718	4.9374
36	1389.03	0.001 707	0.0914	0.0931	352.1	1118.7	1470.8	1.2969	3.6192	4.9161
38	1469.92	0.001 716	0.0863	0.0880	361.9	1109.7	1471.5	1.3281	3.5669	4.8950
40	1554.33	0.001 726	0.0815	0.0833	371.7	1100.5	1472.2	1.3591	3.5148	4.8740
42	1642.35	0.001 735	0.0771	0.0788	381.6	1091.2	1472.8	1.3901	3.4630	4.8530
44	1734.09	0.001 745	0.0728	0.0746	391.5	1081.7	1473.2	1.4209	3.4112	4.8322
46	1829.65	0.001 756	0.0689	0.0707	401.5	1072.0	1473.5	1.4518	3.3595	4.8113
48	1929.13	0.001 766	0.0652	0.0669	411.5	1062.2	1473.7	1.4826	3.3079	4.7905
50	2032.62	0.001 777	0.0617	0.0635	421.7	1052.0	1473.7	1.5135	3.2561	4.7696

Source: Reprinted from Van Wylen, G. J., Sonntag, R. E., 1986. Fundamentals of Classical Thermodynamics, third ed. Wiley, New York. Reprinted by permission of John Wiley & Sons, Inc.

Table C.6a Superheated Ammonia Vapor (English Units)

Abs. Press. psia (Sat. Temp.)		0	20	40	60	80	100	120	140	160	180	200	220
							Temperature, °F						
10 (−41.34)	v	28.58	29.90	31.20	32.49	33.78	35.07	36.35	37.62	38.90	40.17	41.45	—
	h	618.9	629.1	639.3	649.5	659.7	670.0	680.3	690.6	701.1	711.6	722.2	—
	s	1.477	1.499	1.520	1.540	1.559	1.578	1.596	1.614	1.631	1.647	1.664	—
15 (−27.29)	v	18.92	19.82	20.70	21.58	22.44	23.31	24.17	25.03	25.88	26.74	27.59	—
	h	617.2	627.8	638.2	648.5	658.9	669.2	679.6	690.0	700.5	711.1	721.7	—
	s	1.427	1.450	1.471	1.491	1.511	1.529	1.548	1.566	1.583	1.599	1.616	—
20 (−16.64)	v	14.09	14.78	15.45	16.12	16.78	17.43	18.08	18.73	19.37	20.02	20.66	21.3
	h	615.5	626.4	637.0	647.5	658.0	668.5	678.9	689.4	700.0	710.6	721.2	732.0
	s	1.391	1.414	1.436	1.456	1.476	1.495	1.513	1.531	1.549	1.565	1.582	1.598
25 (−7.96)	v	11.19	11.75	12.30	12.84	13.37	13.90	14.43	14.95	15.47	15.99	16.50	17.02
	h	613.8	625.0	635.8	646.5	657.1	667.7	678.2	688.8	699.4	710.1	720.8	731.6
	s	1.362	1.386	1.408	1.429	1.449	1.468	1.486	1.504	1.522	1.539	1.555	1.571
30 (−.57)	v	9.25	9.731	10.20	10.65	11.10	11.55	11.99	12.43	12.87	13.30	13.73	14.16
	h	611.9	623.5	634.6	645.5	656.2	666.9	677.5	688.2	698.8	709.6	720.3	731.1
	s	1.337	1.362	1.385	1.406	1.426	1.446	1.464	1.482	1.500	1.517	1.533	1.550
35 (5.89)	v	—	8.287	8.695	9.093	9.484	9.869	10.25	10.63	11.00	11.38	11.75	12.12
	h	—	622.0	633.4	644.4	655.3	666.1	676.8	687.6	698.3	709.1	719.9	730.7
	s	—	1.341	1.365	1.386	1.407	1.427	1.445	1.464	1.481	1.498	1.515	1.531
40 (11.66)	v	—	7.203	7.568	7.922	8.268	8.609	8.945	9.278	9.609	9.938	10.27	10.59
	h	—	620.4	632.1	643.4	654.4	665.3	676.1	686.9	697.7	708.5	719.4	730.3
	s	—	1.323	1.347	1.369	1.390	1.410	1.429	1.447	1.465	1.482	1.499	1.515
45 (16.87)	v	—	6.363	6.694	7.014	7.326	7.632	7.934	8.232	8.528	8.822	9.115	9.406
	h	—	618.8	630.8	642.3	653.5	664.6	675.5	686.3	697.2	708.0	718.9	729.9
	s	—	1.307	1.331	1.354	1.375	1.395	1.414	1.433	1.450	1.468	1.485	1.501
50 (21.67)	v	—	—	5.988	6.280	6.564	6.843	7.117	7.387	7.655	7.921	8.185	8.448
	h	—	—	629.5	641.2	652.6	663.7	674.7	685.7	696.6	707.5	718.5	729.4
	s	—	—	1.317	1.340	1.361	1.382	1.401	1.420	1.437	1.455	1.472	1.488
60 (30.21)	v	—	—	4.933	5.184	5.428	5.665	5.897	6.126	6.352	6.576	6.798	7.019
	h	—	—	626.8	639.0	650.7	662.1	673.3	684.4	695.5	706.5	717.5	728.6
	s	—	—	1.291	1.315	1.337	1.358	1.378	1.397	1.415	1.432	1.449	1.466

Abs. Press. psia (Sat. Temp.)		Temperature, °F											
		60	80	100	120	140	160	180	200	240	280	320	360
70 (37.7)	v	4.401	4.615	4.822	5.025	5.224	5.420	5.615	5.807	6.187	6.563	—	—
	h	636.6	648.7	660.4	671.8	683.1	694.3	705.5	716.6	738.9	761.4	—	—
	s	1.294	1.317	1.338	1.358	1.377	1.395	1.413	1.430	1.463	1.494	—	—
80 (44.4)	v	3.812	4.005	4.190	4.371	4.548	4.722	4.893	5.063	5.398	5.73	—	—
	h	634.3	646.7	658.7	670.4	681.8	693.2	704.4	715.6	738.1	760.7	—	—
	s	1.275	1.298	1.320	1.340	1.360	1.378	1.396	1.414	1.447	1.478	—	—
90 (50.47)	v	3.353	3.529	3.698	3.862	4.021	4.178	4.332	4.484	4.785	5.081	—	—
	h	631.8	644.7	657.0	668.9	680.5	692.0	703.4	714.7	737.3	760.0	—	—
	s	1.257	1.281	1.304	1.325	1.344	1.363	1.381	1.400	1.432	1.464	—	—
100 (56.05)	v	2.985	3.149	3.304	3.454	3.600	3.743	3.883	4.021	4.294	4.562	—	—
	h	629.3	642.6	655.2	667.3	679.2	690.8	702.3	713.7	736.5	759.4	—	—
	s	1.241	1.266	1.289	1.310	1.331	1.349	1.368	1.385	1.419	1.451	—	—
140 (74.79)	v	—	2.166	2.288	2.404	2.515	2.622	2.727	2.830	3.030	3.227	3.420	—
	h	—	633.8	647.8	661.1	673.7	686.0	698.0	709.9	733.3	756.7	780.0	—
	s	—	1.214	1.240	1.263	1.284	1.305	1.324	1.342	1.376	1.409	1.440	—
180 (89.78)	v	—	—	1.720	1.818	1.910	1.999	2.084	2.167	2.328	2.484	2.637	—
	h	—	—	639.9	654.4	668.0	681.0	693.6	705.9	730.1	753.9	777.7	—
	s	—	—	1.999	1.225	1.248	1.269	1.289	1.308	1.344	1.377	1.408	—
220 (102.42)	v	—	—	—	1.443	1.525	1.601	1.675	1.745	1.881	2.012	2.140	2.265
	h	—	—	—	647.3	662.0	675.8	689.1	701.9	726.8	751.1	775.3	779.5
	s	—	—	—	1.192	1.217	1.239	1.260	1.280	1.317	1.351	1.383	1.413
240 (108.09)	v	—	—	—	1.302	1.380	1.452	1.521	1.587	1.741	1.835	1.954	2.069
	h	—	—	—	643.5	658.8	673.1	686.7	699.8	725.1	749.8	774.1	798.4
	s	—	—	—	1.176	1.203	1.226	1.248	1.268	1.305	1.339	1.371	1.402
260 (113.42)	v	—	—	—	1.182	1.257	1.326	1.391	1.453	1.572	1.686	1.796	1.904
	h	—	—	—	639.5	655.6	670.4	684.4	697.7	723.4	748.4	772.9	797.4
	s	—	—	—	1.162	1.189	1.213	1.235	1.256	1.294	1.329	1.361	1.391
280 (118.45)	v	—	—	—	1.078	1.151	1.217	1.279	1.339	1.451	1.558	1.661	1.762
	h	—	—	—	635.4	652.2	667.6	681.9	695.6	721.8	747.0	771.7	796.3
	s	—	—	—	1.147	1.176	1.201	1.224	1.245	1.283	1.318	1.351	1.382

Note: Units are v – ft³/lbm, h – Btu/lbm, s – Btu/(lbm·R).
Source: Reprinted from Van Wylen, G. J., Sonntag, R. E., 1986. Fundamentals of Classical Thermodynamics, third ed. Wiley, New York. Reprinted by permission of John Wiley & Sons, Inc.

Table C.6b Superheated Ammonia Vapor (Metric Units)

Abs. Press. kPa (Sat. Temp.) °C		Temperature, °C											
		−20	−10	0	10	20	30	40	50	60	70	80	100
50 (−46.54)	v	2.4474	2.5481	2.6482	2.7479	2.8473	2.9464	3.0453	3.1441	3.2427	3.3413	3.4397	—
	h	1435.8	1457.0	1478.1	1499.2	1520.4	1541.7	1563.0	1584.5	1606.1	1627.8	1649.7	—
	s	6.3256	6.4077	6.4865	6.5625	6.6360	6.7073	6.7766	6.8441	6.9099	6.9743	7.0372	—
75 (−39.18)	v	1.6233	1.6915	1.7591	1.8263	1.8932	1.9597	2.0261	2.0923	2.1584	2.2244	2.2903	—
	h	1433.0	1454.7	1476.1	1497.5	1518.9	1540.3	1561.8	1583.4	1605.1	1626.9	1648.9	—
	s	6.1190	6.2028	6.2828	6.3597	6.4339	6.5058	6.5756	6.6434	6.7096	6.7742	6.8373	—
100 (−33.61)	v	1.2110	1.2631	1.3145	1.3654	1.4160	1.4664	1.5165	1.5664	1.6163	1.6659	1.7155	1.8145
	h	1430.1	1452.2	1474.1	1495.7	1517.3	1538.9	1560.5	1582.2	1604.1	1626.0	1648.0	1692.6
	s	5.9695	6.0552	6.1366	6.2144	6.2894	6.3618	6.4321	6.5003	6.5668	6.6316	6.6950	6.8177
125 (−29.08)	v	0.9635	1.0059	1.0476	1.0889	1.1297	1.1703	1.2107	1.2509	1.2909	1.3309	1.3707	1.4501
	h	1427.2	1449.8	1472.0	1493.9	1515.7	1537.5	1559.3	1581.1	1603.0	1625.0	1647.2	1691.8
	s	5.8512	5.9389	6.0217	6.1006	6.1763	6.2494	6.3201	6.3887	6.4555	6.5206	6.5842	6.7072
150 (−25.23)	v	0.7984	0.8344	0.8697	0.9045	0.9388	0.9729	1.0068	1.0405	1.0740	1.1074	1.1408	1.2072
	h	1424.1	1447.3	1469.8	1492.1	1514.1	1536.1	1558.0	1580.0	1602.0	1624.1	1646.3	1691.1
	s	5.7526	5.8424	5.9266	6.0066	6.0831	6.1568	6.2280	6.2970	6.3641	6.4295	6.4933	6.6167
200 (−18.86)	v	—	0.6199	0.6471	0.6738	0.7001	0.7261	0.7519	0.7774	0.8029	0.8282	0.8533	0.9035
	h	—	1442.0	1465.5	1488.4	1510.9	1533.2	1555.5	1577.7	1599.9	1622.2	1644.6	1689.6
	s	—	5.6863	5.7737	5.8559	5.9342	6.0091	6.0813	6.1512	6.2189	6.2849	6.3491	6.4732
250 (−13.67)	v	—	0.4910	0.5135	0.5354	0.5568	0.5780	0.5989	0.6196	0.6401	0.6605	0.6809	0.7212
	h	—	1436.6	1461.0	1484.5	1507.6	1530.3	1552.9	1575.4	1597.8	1620.3	1642.8	1688.2
	s	—	5.5609	5.6517	5.7365	5.8165	5.8928	5.9661	6.0368	6.1052	6.1717	6.2365	6.3613
300 (−9.23)	v	—	—	0.4243	0.4430	0.4613	0.4792	0.4968	0.5143	0.5316	0.5488	0.5658	0.5997
	h	—	—	1456.3	1480.6	1504.2	1527.4	1550.3	1573.0	1595.7	1618.4	1641.1	1686.7
	s	—	—	5.5493	5.6366	5.7186	5.7963	5.8707	5.9423	6.0114	6.0785	6.1437	6.2693
350 (−5.35)	v	—	—	0.3605	0.3770	0.3929	0.4086	0.4239	0.4391	0.4541	0.4689	0.4837	0.5129
	h	—	—	1451.5	1476.5	1500.7	1524.4	1547.6	1570.7	1593.6	1616.5	1639.3	1685.2
	s	—	—	5.4600	5.5502	5.6342	5.7135	5.7890	5.8615	5.9314	5.9990	6.0647	6.1910
400 (−1.89)	v	—	—	0.3125	0.3274	0.3417	0.3556	0.3692	0.3826	0.3959	0.4090	0.4220	0.4478
	h	—	—	1446.5	1472.4	1497.2	1521.3	1544.9	1568.3	1591.5	1614.5	1637.6	1683.7
	s	—	—	5.3803	5.4735	5.5597	5.6405	5.7173	5.7907	5.8613	5.9296	5.9957	6.1228
450 (1.26)	v	—	—	0.2752	0.2887	0.3017	0.3143	0.3266	0.3387	0.3506	0.3624	0.3740	0.3971
	h	—	—	1441.3	1468.1	1493.6	1518.2	1542.2	1565.9	1589.3	1612.6	1635.8	1682.2
	s	—	—	5.3078	5.4042	5.4926	5.5752	5.6532	5.7275	5.7989	5.8678	5.9345	6.0623

Abs. press. kPa (Sat. temp.) °C		Temperature, °C											
		20	30	40	50	60	70	80	100	120	140	160	180
500 (4.14)	v	0.2698	0.2813	0.2926	0.3036	0.3144	0.3251	0.3357	0.3565	0.3771	0.3975	—	—
	h	1489.9	1515.0	1539.5	1563.4	1587.1	1610.6	1634.0	1680.7	1727.5	1774.7	—	—
	s	5.4314	5.5157	5.5950	5.6704	5.7425	5.8120	5.8793	6.0079	6.1301	6.2472	—	—
600 (9.29)	v	0.2217	0.2317	0.2414	0.2508	0.2600	0.2691	0.2781	0.2957	0.3130	0.3302	—	—
	h	1482.4	1508.6	1533.8	1558.5	1582.7	1606.6	1630.4	1677.7	1724.9	1772.4	—	—
	s	5.3222	5.4102	5.4923	5.5697	5.6436	5.7144	5.7826	5.9129	6.0363	6.1541	—	—
700 (13.81)	v	0.1874	0.1963	0.2048	0.2131	0.2212	0.2291	0.2369	0.2522	0.2672	0.2821	—	—
	h	1474.5	1501.9	1528.1	1553.4	1578.2	1602.6	1626.8	1674.6	1722.4	1770.2	—	—
	s	5.2259	5.3179	5.4029	5.4826	5.5582	5.6303	5.6997	5.8316	5.9562	6.0749	—	—
800 (17.86)	v	0.1615	0.1696	0.1773	0.1848	0.1920	0.1991	0.2060	0.2196	0.2329	0.2459	0.2589	—
	h	1466.3	1495.0	1522.2	1548.3	1573.7	1598.6	1623.1	1671.6	1719.8	1768.0	1816.4	—
	s	5.1387	5.2351	5.3232	5.4053	5.4827	5.5562	5.6268	5.7603	5.8861	6.0057	6.1202	—
900 (21.54)	v	—	0.1488	0.1559	0.1627	0.1693	0.1757	0.1820	0.1942	0.2061	0.2178	0.2294	—
	h	—	1488.0	1516.2	1543.0	1569.1	1594.4	1619.4	1668.5	1717.1	1765.7	1814.4	—
	s	—	5.1593	5.2508	5.3354	5.4147	5.4897	5.5614	5.6968	5.8237	5.9442	6.0594	—
1000 (24.91)	v	—	0.1321	0.1388	0.1450	0.1511	0.1570	0.1627	0.1739	0.1847	0.1954	0.2058	0.2162
	h	—	1480.6	1510.0	1537.7	1564.4	1590.3	1615.6	1665.4	1714.5	1763.4	1812.4	1861.7
	s	—	5.0889	5.1840	5.2713	5.3525	5.4292	5.5021	5.6392	5.7674	5.8888	6.0047	6.1159
1200 (30.96)	v	—	—	0.1129	0.1185	0.1238	0.1289	0.1338	0.1434	0.1526	0.1616	0.1705	0.1792
	h	—	—	1497.1	1526.6	1554.7	1581.7	1608.0	1659.2	1709.2	1758.9	1808.5	1858.2
	s	—	—	5.0629	5.1560	5.2416	5.3215	5.3970	5.5379	5.6687	5.7919	5.9091	6.0214
1400 (36.28)	v	—	—	0.0944	0.0995	0.1042	0.1088	0.1132	0.1216	0.1297	0.1376	0.1452	0.1528
	h	—	—	1483.4	1515.1	1544.7	1573.0	1600.2	1652.8	1703.9	1754.3	1804.5	1854.7
	s	—	—	4.9534	5.0530	5.1434	5.2270	5.3053	5.4501	5.5836	5.7087	5.8273	5.9406
1600 (41.05)	v	—	—	—	0.0851	0.0895	0.0937	0.0977	0.1053	0.1125	0.1195	0.1263	0.1330
	h	—	—	—	1502.9	1534.4	1564.0	1592.3	1646.4	1698.5	1749.7	1800.5	1851.2
	s	—	—	—	4.9584	5.0543	5.1419	5.2232	5.3722	5.5084	5.6355	5.7555	5.8699
1800 (45.39)	v	—	—	—	0.0739	0.0781	0.0820	0.0856	0.0926	0.0992	0.1055	0.1116	0.1177
	h	—	—	—	1490.0	1523.5	1554.6	1584.1	1639.8	1693.1	1745.1	1796.5	1847.7
	s	—	—	—	4.8693	4.9715	5.0635	5.1482	5.3018	5.4409	5.5699	5.6914	5.8069
2000 (49.38)	v	—	—	—	0.0648	0.0688	0.0725	0.0760	0.0824	0.0885	0.0943	0.0999	0.1054
	h	—	—	—	1476.1	1512.0	1544.9	1575.6	1633.2	1687.6	1740.3	1792.4	1844.1
	s	—	—	—	4.7834	4.8930	4.9902	5.0786	5.2371	5.3793	5.5104	5.6333	5.7499

Note: Units are v – m³/kg, h – kJ/kg, s – kJ/(kg · K).
Source: Reprinted from Van Wylen, G. J., Sonntag, R. E., 1986. Fundamentals of Classical Thermodynamics, third ed. Wiley, New York. Reprinted by permission of John Wiley & Sons, Inc.

Table C.7c Saturated Refrigerant-134a Temperature Table (Metric Units)

Temp. °C T	Press. MPa P_{sat}	Specific Volume m³/kg		Internal Energy kJ/kg		Enthalpy kJ/kg			Entropy kJ/(kg·K)	
		Sat. Liquid v_f	Sat. Vapor v_g	Sat. Liquid u_f	Sat. Vapor u_g	Sat. Liquid h_f	Evap. h_{fg}	Sat. Vapor h_g	Sat. Liquid s_f	Sat. Vapor s_g
−40	0.051 64	0.000 705 5	0.3569	−0.04	204.45	0.00	222.88	222.88	0.0000	0.9560
−36	0.063 32	0.000 711 3	0.2947	4.68	206.73	4.73	220.67	225.40	0.0201	0.9506
−32	0.077 04	0.000 717 2	0.2451	9.47	209.01	9.52	218.37	227.90	0.0401	0.9456
−28	0.093 05	0.000 723 3	0.2052	14.31	211.29	14.37	216.01	230.38	0.0600	0.9411
−26	0.101 99	0.000 726 5	0.1882	16.75	212.43	16.82	214.80	231.62	0.0699	0.9390
−24	0.111 60	0.000 729 6	0.1728	19.21	213.57	19.29	213.57	232.85	0.0798	0.9370
−22	0.121 92	0.000 732 8	0.1590	21.68	214.70	21.77	212.32	234.08	0.0897	0.9351
−20	0.132 99	0.000 736 1	0.1464	24.17	215.84	24.26	211.05	235.31	0.0996	0.9332
−18	0.144 83	0.000 739 5	0.1350	26.67	216.97	26.77	209.76	236.53	0.1094	0.9315
−16	0.157 48	0.000 742 8	0.1247	29.18	218.10	29.30	208.45	237.74	0.1192	0.9298
−12	0.185 40	0.000 749 8	0.1068	34.25	220.36	34.39	205.77	240.15	0.1388	0.9267
−8	0.217 04	0.000 756 9	0.0919	39.38	222.60	39.54	203.00	242.54	0.1583	0.9239
−4	0.252 74	0.000 764 4	0.0794	44.56	224.84	44.75	200.15	244.90	0.1777	0.9213
0	0.292 82	0.000 772 1	0.0689	49.79	227.06	50.02	197.21	247.23	0.1970	0.9190
4	0.337 65	0.000 780 1	0.0600	55.08	229.27	55.35	194.19	249.53	0.2162	0.9169
8	0.387 56	0.000 788 4	0.0525	60.43	231.46	60.73	191.07	251.80	0.2354	0.9150
12	0.442 94	0.000 797 1	0.0460	65.83	233.63	66.18	187.85	254.03	0.2545	0.9132
16	0.504 16	0.000 806 2	0.0405	71.29	235.78	71.69	184.52	256.22	0.2735	0.9116
20	0.571 60	0.000 815 7	0.0358	76.80	237.91	77.26	181.09	258.36	0.2924	0.9102
24	0.645 66	0.000 825 7	0.0317	82.37	240.01	82.90	177.55	260.45	0.3113	0.9089
26	0.685 30	0.000 830 9	0.0298	85.18	241.05	85.75	175.73	261.48	0.3208	0.9082
28	0.726 75	0.000 836 2	0.0281	88.00	242.08	88.61	173.89	262.50	0.3302	0.9076
30	0.770 06	0.000 841 7	0.0265	90.84	243.10	91.49	172.00	263.50	0.3396	0.9070
32	0.815 28	0.000 847 3	0.0250	93.70	244.12	94.39	170.09	264.48	0.3490	0.9064
34	0.862 47	0.000 853 0	0.0236	96.58	245.12	97.31	168.14	265.45	0.3584	0.9058
36	0.911 68	0.000 859 0	0.0223	99.47	246.11	100.25	166.15	266.40	0.3678	0.9053
38	0.962 98	0.000 865 1	0.0210	102.38	247.09	103.21	164.12	267.33	0.3772	0.9047
40	1.016 4	0.000 871 4	0.0199	105.30	248.06	106.19	162.05	268.24	0.3866	0.9041
42	1.072 0	0.000 878 0	0.0188	108.25	249.02	109.19	159.94	269.14	0.3960	0.9035
44	1.129 9	0.000 884 7	0.0177	111.22	249.96	112.22	157.79	270.01	0.4054	0.9030
48	1.252 6	0.000 898 9	0.0159	117.22	251.79	118.35	153.33	271.68	0.4243	0.9017
52	1.385 1	0.000 914 2	0.0142	123.31	253.55	124.58	148.66	273.24	0.4432	0.9004
56	1.527 8	0.000 930 8	0.0127	129.51	255.23	130.93	143.75	274.68	0.4622	0.8990
60	1.681 3	0.000 948 8	0.0114	135.82	256.81	137.42	138.57	275.99	0.4814	0.8973
70	2.116 2	0.001 002 7	0.0086	152.22	260.15	154.34	124.08	278.43	0.5302	0.8918
80	2.632 4	0.001 076 6	0.0064	169.88	262.14	172.71	106.41	279.12	0.5814	0.8827
90	3.243 5	0.001 194 9	0.0046	189.82	261.34	193.69	82.63	276.32	0.6380	0.8655
100	3.974 2	0.001 544 3	0.0027	218.60	248.49	224.74	34.40	259.13	0.7196	0.8117

Source: Adapted from Moran, M. J., Shapiro, H. N., 1992. Fundamentals of Engineering Thermodynamics, second ed. Wiley, New York, pp. 710–715. Originally based on equations from Wilson, D. P., Basu, R. S., 1988. Thermodynamic properties, of a new stratospherically safe working fluid—refrigerant 134a. ASHRAE Trans. 94 (Pt. 2), 2095–2118.

Table C.7d Saturated Refrigerant-134a Pressure Table (Metric Units)

Press. MPa P	Temp. °C T_{sat}	Specific Volume m³/kg		Internal Energy kJ/kg		Enthalpy kJ/kg			Entropy kJ/(kg·K)	
		Sat. Liquid v_f	Sat. Vapor v_g	Sat. Liquid u_f	Sat. Vapor u_g	Sat. Liquid h_f	Evap. h_{fg}	Sat. Vapor h_g	Sat. Liquid s_f	Sat. Vapor s_g
0.06	−37.07	0.000 709 7	0.3100	3.41	206.12	3.46	221.27	224.72	0.0147	0.9520
0.08	−31.21	0.000 718 4	0.2366	10.41	209.46	10.47	217.92	228.39	0.0440	0.9447
0.10	−26.43	0.000 725 8	0.1917	16.22	212.18	16.29	215.06	231.35	0.0678	0.9395
0.12	−22.36	0.000 732 3	0.1614	21.23	214.50	21.32	212.54	233.86	0.0879	0.9354
0.14	−18.80	0.000 738 1	0.1395	25.66	216.50	25.77	210.27	236.04	0.1055	0.9322
0.16	−15.62	0.000 743 5	0.1229	29.66	218.32	29.78	208.18	237.97	0.1211	0.9295
0.18	−12.73	0.000 748 5	0.1098	33.31	219.94	33.45	206.26	239.71	0.1352	0.9273
0.20	−10.09	0.000 753 2	0.0993	36.69	221.43	36.84	204.46	241.30	0.1481	0.9253
0.24	−5.37	0.000 761 8	0.0834	42.77	224.07	42.95	201.14	244.09	0.1710	0.9222
0.28	−1.23	0.000 769 7	0.0719	48.18	226.38	48.39	198.13	246.52	0.1911	0.9197
0.32	2.48	0.000 777 0	0.0632	53.06	228.43	53.31	195.35	248.66	0.2089	0.9177
0.36	5.84	0.000 783 9	0.0564	57.54	230.28	57.82	192.76	250.58	0.2251	0.9160
0.4	8.93	0.000 790 4	0.0509	61.69	231.97	62.00	190.32	252.32	0.2399	0.9145
0.5	15.74	0.000 805 6	0.0409	70.93	235.64	71.33	184.74	256.07	0.2723	0.9117
0.6	21.58	0.000 819 6	0.0341	78.99	238.74	79.48	179.71	259.19	0.2999	0.9097
0.7	26.72	0.000 832 8	0.0292	86.19	241.42	86.78	175.07	261.85	0.3242	0.9080
0.8	31.33	0.000 845 4	0.0255	92.75	243.78	93.42	170.73	264.15	0.3459	0.9066
0.9	35.53	0.000 857 6	0.0226	98.79	245.88	99.56	166.62	266.18	0.3656	0.9054
1.0	39.39	0.000 869 5	0.0202	104.42	247.77	105.29	162.68	267.97	0.3838	0.9043
1.2	46.32	0.000 892 8	0.0166	114.69	251.03	115.76	155.23	270.99	0.4164	0.9023
1.4	52.43	0.000 915 9	0.0140	123.98	253.74	125.26	148.14	273.40	0.4453	0.9003
1.6	57.92	0.000 939 2	0.0121	132.52	256.00	134.02	141.31	275.33	0.4714	0.8982
1.8	62.91	0.000 963 1	0.0105	140.49	257.88	142.22	134.60	276.83	0.4954	0.8959
2.0	67.49	0.000 987 8	0.0093	148.02	259.41	149.99	127.95	277.94	0.5178	0.8934
2.5	77.59	0.001 056 2	0.0069	165.48	261.84	168.12	111.06	279.17	0.5687	0.8854
3.0	86.22	0.001 141 6	0.0053	181.88	262.16	185.30	92.71	278.01	0.6156	0.8735

Source: Adapted from Moran, M. J., Shapiro, H. N., 1992. Fundamentals of Engineering Thermodynamics, second ed. Wiley, New York, pp. 710–715. Originally based on equations from Wilson, D. P., Basu, R. S., 1988. Thermodynamic properties, of a new stratospherically safe working fluid—refrigerant 134a. ASHRAE Trans. 94 (Pt. 2), 2095–2118.

Table C.8a Superheated Refrigerant-134a Vapor (English Units)

T °F	v ft³/lbm	u Btu/lbm	h Btu/lbm	s Btu/(lbm · R)	v ft³/lbm	u Btu/lbm	h Btu/lbm	s Btu/(lbm · R)
	P = 10 psia (T_sat = −29.71°F)				P = 15 psia (T_sat = −14.25°F)			
Sat.	4.3581	89.30	97.37	0.2265	2.9747	91.40	99.66	0.2242
−20	4.4718	90.89	99.17	0.2307	–	–	–	–
0	4.7026	94.24	102.94	0.2391	3.0893	93.84	102.42	0.2303
20	4.9297	97.67	106.79	0.2472	3.2468	97.33	106.34	0.2386
40	5.1539	101.19	110.72	0.2553	3.4012	100.89	110.33	0.2468
60	5.3758	104.80	114.74	0.2632	3.5533	104.54	114.40	0.2548
80	5.5959	108.50	118.85	0.2709	3.7034	108.28	118.56	0.2626
100	5.8145	112.29	123.05	0.2786	3.8520	112.10	122.79	0.2703
120	6.0318	116.18	127.34	0.2861	3.9993	116.01	127.11	0.2779
140	6.2482	120.16	131.72	0.2935	4.1456	120.00	131.51	0.2854
160	6.4638	124.23	136.19	0.3009	4.2911	124.09	136.00	0.2927
180	6.6786	128.38	140.74	0.3081	4.4359	128.26	140.57	0.3000
200	6.8929	132.63	145.39	0.3152	4.5801	132.52	145.23	0.3072
	P = 20 psia (T_sat = −2.48°F)				P = 30 psia (T_sat = −15.38°F)			
Sat.	2.2661	93.00	101.39	0.2227	1.5408	95.40	103.96	0.2209
0	2.2816	93.43	101.88	0.2238	–	–	–	–
20	2.4046	96.98	105.88	0.2323	1.5611	96.26	104.92	0.2229
40	2.5244	100.59	109.94	0.2406	1.6465	99.98	109.12	0.2315
60	2.6416	104.28	114.06	0.2487	1.7293	103.75	113.35	0.2398
80	2.7569	108.05	118.25	0.2566	1.8098	107.59	117.63	0.2478
100	2.8705	111.90	122.52	0.2644	1.8887	111.49	121.98	0.2558
120	2.9829	115.83	126.87	0.2720	1.9662	115.47	126.39	0.2635
140	3.0942	119.85	131.30	0.2795	2.0426	119.53	130.87	0.2711
160	3.2047	123.95	135.81	0.2869	2.1181	123.66	135.42	0.2786
180	3.3144	128.13	140.40	0.2922	2.1929	127.88	140.05	0.2859
200	3.4236	132.40	145.07	0.3014	2.2671	132.17	144.76	0.2932
220	3.5323	136.76	149.83	0.3085	2.3407	136.55	149.54	0.3003
	P = 40 psia (T_sat = 29.04°F)				P = 50 psia (T_sat = 40.27°F)			
Sat.	1.1692	97.23	105.88	0.2197	0.9422	98.71	107.43	0.2189
40	1.2065	99.33	108.26	0.2245	–	–	–	–
60	1.2723	103.20	112.62	0.2331	0.9974	102.62	111.85	0.2276
80	1.3357	107.11	117.00	0.2414	1.0508	106.62	116.34	0.2361
100	1.3973	111.08	121.42	0.2494	1.1022	110.65	120.85	0.2443
120	1.4575	115.11	125.90	0.2573	1.1520	114.74	125.39	0.2523
140	1.5165	119.21	130.43	0.2650	1.2007	118.88	129.99	0.2601
160	1.5746	123.38	135.03	0.2725	1.2484	123.08	134.64	0.2677
180	1.6319	127.62	139.70	0.2799	1.2953	127.36	139.34	0.2752
200	1.6887	131.94	144.44	0.2872	1.3415	131.71	144.12	0.2825
220	1.7449	136.34	149.25	0.2944	1.3873	136.12	148.96	0.2897
240	1.8006	140.81	154.14	0.3015	1.4326	140.61	153.87	0.2969
260	1.8561	145.36	159.10	0.3085	1.4775	145.18	158.85	0.3039
280	1.9112	149.98	164.13	0.3154	1.5221	149.82	163.90	0.3108

Table C.8a Superheated Refrigerant-134a Vapor (English Units)   *continued*

T °F	v ft³/lbm	u Btu/lbm	h Btu/lbm	s Btu/ (lbm · R)	v ft³/lbm	u Btu/lbm	h Btu/lbm	s Btu/ (lbm · R)
	P = 60 psia (T_{sat} = 49.89°F)				P = 70 psia (T_{sat} = 58.35°F)			
Sat.	0.7887	99.96	108.72	0.2183	0.6778	101.05	109.83	0.2179
60	0.8135	102.03	111.06	0.2229	0.6814	101.40	110.23	0.2186
80	0.8604	106.11	115.66	0.2316	0.7239	105.58	114.96	0.2276
100	0.9051	110.21	120.26	0.2399	0.7640	109.76	119.66	0.2361
120	0.9482	114.35	124.88	0.2480	0.8023	113.96	124.36	0.2444
140	0.9900	118.54	129.53	0.2559	0.8393	118.20	129.07	0.2524
160	1.0308	122.79	134.23	0.2636	0.8752	122.49	133.82	0.2601
180	1.0707	127.10	138.98	0.2712	0.9103	126.83	138.62	0.2678
200	1.1100	131.47	143.79	0.2786	0.9446	131.23	143.46	0.2752
220	1.1488	135.91	148.66	0.2859	0.9784	135.69	148.36	0.2825
240	1.1871	140.42	153.60	0.2930	1.0118	140.22	153.33	0.2897
260	1.2251	145.00	158.60	0.3001	1.0448	144.82	158.35	0.2968
280	1.2627	149.65	163.67	0.3070	1.0774	149.48	163.44	0.3038
300	1.3001	154.38	168.81	0.3139	1.1098	154.22	168.60	0.3107
	P = 80 psia (T_{sat} = 65.93°F)				P = 90 psia (T_{sat} = 72.83°F)			
Sat.	0.5938	102.02	110.81	0.2175	0.5278	102.89	111.68	0.2172
80	0.6211	105.03	114.23	0.2239	0.5408	104.46	113.47	0.2205
100	0.6579	109.30	119.04	0.2327	0.5751	108.82	118.39	0.2295
120	0.6927	113.56	123.82	0.2411	0.6073	113.15	123.27	0.2380
140	0.7261	117.85	128.60	0.2492	0.6380	117.50	128.12	0.2463
160	0.7584	122.18	133.41	0.2570	0.6675	121.87	132.98	0.2542
180	0.7898	126.55	138.25	0.2647	0.6961	126.28	137.87	0.2620
200	0.8205	130.98	143.13	0.2722	0.7239	130.73	142.79	0.2696
220	0.8506	135.47	148.06	0.2796	0.7512	135.25	147.76	0.2770
240	0.8803	140.02	153.05	0.2868	0.7779	139.82	152.77	0.2843
260	0.9095	144.63	158.10	0.2940	0.8043	144.45	157.84	0.2914
280	0.9384	149.32	163.21	0.3010	0.8303	149.15	162.97	0.2984
300	0.9671	154.06	168.38	0.3079	0.8561	153.91	168.16	0.3054
320	0.9955	158.88	173.62	0.3147	0.8816	158.73	173.42	0.3122
	P = 100 psia (T_{sat} = 79.17°F)				P = 120 psia (T_{sat} = 90.54°F)			
Sat.	0.4747	103.68	112.46	0.2169	0.3941	105.06	113.82	0.2165
80	0.4761	103.87	112.68	0.2173	–	–	–	–
100	0.5086	108.32	117.73	0.2265	0.4080	107.26	116.32	0.2210
120	0.5388	112.73	122.70	0.2352	0.4355	111.84	121.52	0.2301
140	0.5674	117.13	127.63	0.2436	0.4610	116.37	126.61	0.2387
160	0.5947	121.55	132.55	0.2517	0.4852	120.89	131.66	0.2470
180	0.6210	125.99	137.49	0.2595	0.5082	125.42	136.70	0.2550
200	0.6466	130.48	142.45	0.2671	0.5305	129.97	141.75	0.2628
220	0.6716	135.02	147.45	0.2746	0.5520	134.56	146.82	0.2704
240	0.6960	139.61	152.49	0.2819	0.5731	139.20	151.92	0.2778
260	0.7201	144.26	157.59	0.2891	0.5937	143.89	157.07	0.2850
280	0.7438	148.98	162.74	0.2962	0.6140	148.63	162.26	0.2921
300	0.7672	153.75	167.95	0.3031	0.6339	153.43	167.51	0.2991
320	0.7904	158.59	173.21	0.3099	0.6537	158.29	172.81	0.3060

(Continued)

Table C.8a Superheated Refrigerant-134a Vapor (English Units) *continued*

T °F	v ft³/lbm	u Btu/lbm	h Btu/lbm	s Btu/ (lbm · R)	v ft³/lbm	u Btu/lbm	h Btu/lbm	s Btu/ (lbm · R)
	$P = 140$ psia ($T_{sat} = 100.56°F$)				$P = 160$ psia ($T_{sat} = 109.55°F$)			
Sat.	0.3358	106.25	114.95	0.2161	0.2916	107.28	115.91	0.2157
120	0.3610	110.90	120.25	0.2254	0.3044	109.88	118.89	0.2209
140	0.3846	115.58	125.24	0.2344	0.3269	114.73	124.41	0.2303
160	0.4066	120.21	130.74	0.2429	0.3474	119.49	129.78	0.2391
180	0.4274	124.82	135.89	0.2511	0.3666	124.20	135.06	0.2475
200	0.4474	129.44	141.03	0.2590	0.3849	128.90	140.29	0.2555
220	0.4666	134.09	146.18	0.2667	0.4023	133.61	145.52	0.2633
240	0.4852	138.77	151.34	0.2742	0.4192	138.34	150.75	0.2709
260	0.5034	143.50	156.54	0.2815	0.4356	143.11	156.00	0.2783
280	0.5212	148.28	161.78	0.2887	0.4516	147.92	161.29	0.2856
300	0.5387	153.11	167.06	0.2957	0.4672	152.78	166.61	0.2927
320	0.5559	157.99	172.39	0.3026	0.4826	157.69	171.98	0.2996
340	0.5730	162.93	177.78	0.3094	0.4978	162.65	177.39	0.3065
360	0.5898	167.93	183.21	0.3162	0.5128	167.67	182.85	0.3132
	$P = 180$ psia ($T_{sat} = 117.74°F$)				$P = 200$ psia ($T_{sat} = 125.28°F$)			
Sat.	0.2569	108.18	116.74	0.2154	0.2288	108.98	117.44	0.2151
120	0.2595	108.77	117.41	0.2166	–	–	–	–
140	0.2814	113.83	123.21	0.2264	0.2446	112.87	121.92	0.2226
160	0.3011	118.74	128.77	0.2355	0.2636	117.94	127.70	0.2321
180	0.3191	123.56	134.19	0.2441	0.2809	122.88	133.28	0.2410
200	0.3361	128.34	139.53	0.2524	0.2970	127.76	138.75	0.2494
220	0.3523	133.11	144.84	0.2603	0.3121	132.60	144.15	0.2575
240	0.3678	137.90	150.15	0.2680	0.3266	137.44	149.53	0.2653
260	0.3828	142.71	155.46	0.2755	0.3405	142.30	154.90	0.2728
280	0.3974	147.55	160.79	0.2828	0.3540	147.18	160.28	0.2802
300	0.4116	152.44	166.15	0.2899	0.3671	152.10	165.69	0.2874
320	0.4256	157.38	171.55	0.2969	0.3799	157.07	171.13	0.2945
340	0.4393	162.36	177.00	0.3038	0.3926	162.07	176.60	0.3014
360	0.4529	167.40	182.49	0.3106	0.4050	167.13	182.12	0.3082
	$P = 300$ psia ($T_{sat} = 156.17°F$)				$P = 400$ psia ($T_{sat} = 179.95°F$)			
Sat.	0.1424	111.72	119.62	0.2132	0.0965	112.77	119.91	0.2102
160	0.1462	112.95	121.07	0.2155	–	–	–	–
180	0.1633	118.93	128.00	0.2265	0.0965	112.79	119.93	0.2102
200	0.1777	124.47	134.34	0.2363	0.1143	120.14	128.60	0.2235
220	0.1905	129.79	140.36	0.2453	0.1275	126.35	135.79	0.2343
240	0.2021	134.99	146.21	0.2537	0.1386	132.12	142.38	0.2438
260	0.2130	140.12	151.95	0.2618	0.1484	137.65	148.64	0.2527
280	0.2234	145.23	157.63	0.2696	0.1575	143.06	154.72	0.2610
300	0.2333	150.33	163.28	0.2772	0.1660	148.39	160.67	0.2689
320	0.2428	155.44	168.92	0.2845	0.1740	153.69	166.57	0.2766
340	0.2521	160.57	174.56	0.2916	0.1816	158.97	172.42	0.2840
360	0.2611	165.74	180.23	0.2986	0.1890	164.26	178.26	0.2912
380	0.2699	170.94	185.92	0.3055	0.1962	169.57	184.09	0.2983
400	0.2786	176.18	191.64	0.3122	0.2032	174.90	189.94	0.3051

Table C.8b Superheated Refrigerant-134a Vapor (Metric Units)

T °C	v m³/kg	u kJ/kg	h kJ/kg	s kJ/(kg·K)	v m³/kg	u kJ/kg	h kJ/kg	s kJ/(kg·K)
	P = 0.06 MPa (T_sat = −37.07°C)				P = 0.10 MPa (T_sat = −26.43°C)			
Sat.	0.31003	206.12	224.72	0.9520	0.19170	212.18	231.35	0.9395
−20	0.33536	217.86	237.98	1.0062	0.19770	216.77	236.54	0.9602
−10	0.34992	224.97	245.96	1.0371	0.20686	224.01	244.70	0.9918
0	0.36433	232.24	254.10	1.0675	0.21587	231.41	252.99	1.0227
10	0.37861	239.69	262.41	1.0973	0.22473	238.96	261.43	1.0531
20	0.39279	247.32	270.89	1.1267	0.23349	246.67	270.02	1.0829
30	0.40688	255.12	279.53	1.1557	0.24216	254.54	278.76	1.1122
40	0.42091	263.10	288.35	1.1844	0.25076	262.58	287.66	1.1411
50	0.43487	271.25	297.34	1.2126	0.25930	270.79	296.72	1.1696
60	0.44879	279.58	306.51	1.2405	0.26779	279.16	305.94	1.1977
70	0.46266	288.08	315.84	1.2681	0.27623	287.70	315.32	1.2254
80	0.47650	296.75	325.34	1.2954	0.28464	296.40	324.87	1.2528
90	0.49031	305.58	335.00	1.3224	0.29302	305.27	334.57	1.2799
	P = 0.14 MPa (T_sat = −18.80°C)				P = 0.18 MPa (T_sat = −12.73°C)			
Sat.	0.13945	216.52	236.04	0.9322	0.10983	219.94	239.71	0.9273
−10	0.14549	223.03	243.40	0.9606	0.11135	222.02	242.06	0.9362
0	0.15219	230.55	251.86	0.9922	0.11678	229.67	250.69	0.9684
10	0.15875	238.21	260.43	1.0230	0.12207	237.44	259.41	0.9998
20	0.16520	246.01	269.13	1.0532	0.12723	245.33	268.23	1.0304
30	0.17155	253.96	277.97	1.0828	0.13230	253.36	277.17	1.0604
40	0.17783	262.06	286.96	1.1120	0.13730	261.53	286.24	1.0898
50	0.18404	270.32	296.09	1.1407	0.14222	269.85	295.45	1.1187
60	0.19020	278.74	305.37	1.1690	0.14710	278.31	304.79	1.1472
70	0.19633	287.32	314.80	1.1969	0.15193	286.93	314.28	1.1753
80	0.20241	296.06	324.39	1.2244	0.15672	295.71	323.92	1.2030
90	0.20846	304.95	334.14	1.2516	0.16148	304.63	333.70	1.2303
100	0.21449	314.01	344.04	1.2785	0.16622	313.72	343.63	1.2573
	P = 0.20 MPa (T_sat = −10.09°C)				P = 0.24 MPa (T_sat = −5.37°C)			
Sat.	0.09933	221.43	241.30	0.9253	0.08343	224.07	244.09	0.9222
−10	0.09938	221.50	241.38	0.9256	–	–	–	–
0	0.10438	229.23	250.10	0.9582	0.08574	228.31	248.89	0.9399
10	0.10922	237.05	258.89	0.9898	0.08993	236.26	257.84	0.9721
20	0.11394	244.99	267.78	1.0206	0.09399	244.30	266.85	1.0034
30	0.11856	253.06	276.77	1.0508	0.09794	252.45	275.95	1.0339
40	0.12311	261.26	285.88	1.0804	0.10181	260.72	285.16	1.0637
50	0.12758	269.61	295.12	1.1094	0.10562	269.12	294.47	1.0930
60	0.13201	278.10	304.50	1.1380	0.10937	277.67	303.91	1.1218
70	0.13639	286.74	314.02	1.1661	0.11307	286.35	313.49	1.1501
80	0.14073	295.53	323.68	1.1939	0.11674	295.18	323.19	1.1780
90	0.14504	304.47	333.48	1.2212	0.12037	304.15	333.04	1.2055
100	0.14932	313.57	343.43	1.2483	0.12398	313.27	343.03	1.2326
	P = 0.28 MPa (T_sat = −1.23°C)				P = 0.32 MPa (T_sat = 2.48°C)			
Sat.	0.07193	226.38	246.52	0.9197	0.06322	228.43	248.66	0.9177
0	0.07240	227.37	247.64	0.9238	–	–	–	–
10	0.07613	235.44	256.76	0.9566	0.06576	234.61	255.65	0.9427
20	0.07972	243.59	265.91	0.9883	0.06901	242.87	264.95	0.9749
30	0.08320	251.83	275.12	1.0192	0.07214	251.19	274.28	1.0062

Table C.8b Superheated Refrigerant-134a Vapor (Metric Units) *continued*

T °C	v m³/kg	u kJ/kg	h kJ/kg	s kJ/(kg·K)	v m³/kg	u kJ/kg	h kJ/kg	s kJ/(kg·K)
40	0.08660	260.17	284.42	1.0494	0.07518	259.61	283.67	1.0367
50	0.08992	268.64	293.81	1.0789	0.07815	268.14	293.15	1.0665
60	0.09319	277.23	303.32	1.1079	0.08106	276.79	302.72	1.0957
70	0.09641	285.96	312.95	1.1364	0.08392	285.56	312.41	1.1243
80	0.09960	294.82	322.71	1.1644	0.08674	294.46	322.22	1.1525
90	0.10275	303.83	332.60	1.1920	0.08953	303.50	332.15	1.1802
100	0.10587	312.98	342.62	1.2193	0.09229	312.68	342.21	1.2076
110	0.10897	322.27	352.78	1.2461	0.09503	322.00	352.40	1.2345
120	0.11205	331.71	363.08	1.2727	0.09774	331.45	362.73	1.2611
	P = 0.40 MPa (*T*$_{sat}$ = 8.93°C)				*P* = 0.50 MPa (*T*$_{sat}$ = 15.74°C)			
Sat.	0.05089	231.97	252.32	0.9145	0.04086	235.64	256.07	0.9117
10	0.05119	232.87	253.35	0.9182	–	–	–	–
20	0.05397	241.37	262.96	0.9515	0.04188	239.40	260.34	0.9264
30	0.05662	249.89	272.54	0.8937	0.04416	248.20	270.28	0.9597
40	0.05917	258.47	282.14	1.0148	0.04633	256.99	280.16	0.9918
50	0.06164	267.13	291.79	1.0452	0.04842	265.83	290.04	1.0229
60	0.06405	275.89	301.51	1.0748	0.05043	274.73	299.95	1.0531
70	0.06641	284.75	311.32	1.1038	0.05240	283.72	309.92	1.0825
80	0.06873	293.73	321.23	1.1322	0.05432	292.80	319.96	1.1114
90	0.07102	302.84	331.25	1.1602	0.05620	302.00	330.10	1.1397
100	0.07327	312.07	341.38	1.1878	0.05805	311.31	340.33	1.1675
110	0.07550	321.44	351.64	1.2149	0.05988	320.74	350.68	1.1949
120	0.07771	330.94	362.03	1.2417	0.06168	330.30	361.14	1.2218
130	0.07991	340.58	372.54	1.2681	0.06347	339.98	371.72	1.2484
140	0.08208	350.35	383.18	1.2941	0.06524	349.79	382.42	1.2746
	P = 0.60 MPa (*T*$_{sat}$ = 21.58°C)				*P* = 0.70 MPa (*T*$_{sat}$ = 26.72°C)			
Sat.	0.03408	238.74	259.19	0.9097	0.02918	241.42	261.85	0.9080
30	0.03581	246.41	267.89	0.9388	0.02979	244.51	265.37	0.9197
40	0.03774	255.45	278.09	0.9719	0.03157	253.83	275.93	0.9539
50	0.03958	264.48	288.23	1.0037	0.03324	263.08	286.35	0.9867
60	0.04134	273.54	298.35	1.0346	0.03482	272.31	296.69	1.0182
70	0.04304	282.66	308.48	1.0645	0.03634	281.57	307.01	1.0487
80	0.04469	291.86	318.67	1.0938	0.03781	290.88	317.35	1.0784
90	0.04631	301.14	328.93	1.1225	0.03924	300.27	327.74	1.1074
100	0.04790	310.53	339.27	1.1505	0.04064	309.74	338.19	1.1358
110	0.04946	320.03	349.70	1.1781	0.04201	319.31	348.71	1.1637
120	0.05099	329.64	360.24	1.2053	0.04335	328.98	359.33	1.1910
130	0.05251	339.38	370.88	1.2320	0.04468	338.76	370.04	1.2179
140	0.05402	349.23	381.64	1.2584	0.04599	348.66	380.86	1.2444
150	0.05550	359.21	392.52	1.2844	0.04729	358.68	391.79	1.2706
160	0.05698	369.32	403.51	1.3100	0.04857	368.82	402.82	1.2963
	P = 0.80 MPa (*T*$_{sat}$ = 31.33°C)				*P* = 0.90 MPa (*T*$_{sat}$ = 35.53°C)			
Sat.	0.02547	243.78	264.15	0.9066	0.02255	245.88	266.18	0.9054
40	0.02691	252.13	273.66	0.9374	0.02325	250.32	271.25	0.9217
50	0.02846	261.62	284.39	0.9711	0.02472	260.09	282.34	0.9566
60	0.02992	271.04	294.98	1.0034	0.02609	269.72	293.21	0.9897
70	0.03131	280.45	305.50	1.0345	0.02738	279.30	303.94	1.0214
80	0.03264	289.89	316.00	1.0647	0.02861	288.87	314.62	1.0521

Table C.8b Superheated Refrigerant-134a Vapor (Metric Units) *continued*

T °C	v m³/kg	u kJ/kg	h kJ/kg	s kJ/(kg·K)	v m³/kg	u kJ/kg	h kJ/kg	s kJ/(kg·K)
90	0.03393	299.37	326.52	1.0940	0.02980	298.46	325.28	1.0819
100	0.03519	308.93	337.08	1.1227	0.03095	308.11	335.96	1.1109
110	0.03642	318.57	347.71	1.1508	0.03207	317.82	346.68	1.1392
120	0.03762	328.31	358.40	1.1784	0.03316	327.62	357.47	1.1670
130	0.03881	338.14	369.19	1.2055	0.03423	337.52	368.33	1.1943
140	0.03997	348.09	380.07	1.2321	0.03529	347.51	379.27	1.2211
150	0.04113	358.15	391.05	1.2584	0.03633	357.61	390.31	1.2475
160	0.04227	368.32	402.14	1.2843	0.03736	367.82	401.44	1.2735
170	0.04340	378.61	413.33	1.3098	0.03838	378.14	412.68	1.2992
180	0.04452	389.02	424.63	1.3351	0.03939	388.57	424.02	1.3245
	P = 1.00 MPa (T_sat = 39.39°C)				**P = 1.20 MPa (T_sat = 46.32°C)**			
Sat.	0.02020	*247.77*	267.97	0.9043	0.01663	251.03	270.99	0.9023
40	0.02029	248.39	268.68	0.9066	–	–	–	–
50	0.02171	258.48	280.19	0.9428	0.01712	254.98	275.52	0.9164
60	0.02301	268.35	291.36	0.9768	0.01835	265.42	287.44	0.9527
70	0.02423	278.11	302.34	1.0093	0.01947	275.59	298.96	0.9868
80	0.02538	287.82	313.20	1.0405	0.02051	285.62	310.24	1.0192
90	0.02649	297.53	324.01	1.0707	0.02150	295.59	321.39	1.0503
100	0.02755	307.27	334.82	1.1000	0.02244	305.54	332.47	1.0804
110	0.02858	317.06	345.65	1.1286	0.02335	315.50	343.52	1.1096
120	0.02959	326.93	356.52	1.1567	0.02423	325.51	354.58	1.1381
130	0.03058	336.88	367.46	1.1841	0.02508	335.58	365.68	1.1660
140	0.03154	346.92	378.46	1.2111	0.02592	345.73	376.83	1.1933
150	0.03250	357.06	389.56	1.2376	0.02674	355.95	388.04	1.2201
160	0.03344	367.31	400.74	1.2638	0.02754	366.27	399.33	1.2465
170	0.03436	377.66	412.02	1.2895	0.02834	376.69	410.70	1.2724
180	0.03528	388.12	423.40	1.3149	0.02912	387.21	422.16	1.2980
	P = 1.40 MPa (T_sat = 52.43°C)				**P = 1.60 MPa (T_sat = 57.92°C)**			
Sat.	0.01405	253.74	273.40	0.9003	0.01208	256.00	275.33	0.8982
60	0.01495	262.17	283.10	0.9297	0.01233	258.48	278.20	0.9069
70	0.01603	272.87	295.31	0.9658	0.01340	269.89	291.33	0.9457
80	0.01701	283.29	307.10	0.9997	0.01435	280.78	303.74	0.9813
90	0.01792	293.55	318.63	1.0319	0.01521	291.39	315.72	1.0148
100	0.01878	303.73	330.02	1.0628	0.01601	301.84	327.46	1.0467
110	0.01960	313.88	341.32	1.0927	0.01677	312.20	339.04	1.0773
120	0.02039	324.05	352.59	1.1218	0.01750	322.53	350.53	1.1069
130	0.02115	334.25	363.86	1.1501	0.01820	332.87	361.99	1.1357
140	0.02189	344.50	375.15	1.1777	0.01887	343.24	373.44	1.1638
150	0.02262	354.82	386.49	1.2048	0.01953	353.66	384.91	1.1912
160	0.02333	365.22	397.89	1.2315	0.02017	364.15	396.43	1.2181
170	0.02403	375.71	409.36	1.2576	0.02080	374.71	407.99	1.2445
180	0.02472	386.29	420.90	1.2834	0.02142	385.35	419.62	1.2704
190	0.02541	396.96	432.53	1.3088	0.02203	396.08	431.33	1.2960
200	0.02608	407.73	444.24	1.3338	0.02263	406.90	443.11	1.3212

Table C.9a Saturated Refrigerant-22, Temperature Table (English Units)

Temp. °F T	Abs. press. pisa p	Specific volume (ft³/lbm)			Internal energy (Btu/lbm)		
		Sat. liq. v_f	Evap. v_{fg}	Sat. vap. v_g	Sat. liq. u_f	Evap. u_{fg}	Sat. vap. u_g
−150	0.2716	0.01018	141.22	141.23	−25.98	106.40	80.42
−140	0.4469	0.01027	88.522	88.532	−23.73	105.09	81.36
−130	0.7106	0.01037	57.346	57.356	−21.46	103.77	83.31
−120	1.0954	0.01046	38.270	38.280	−10.19	102.45	83.26
−110	1.6417	0.01056	26.231	26.242	−16.89	101.11	84.22
−100	2.3989	0.01066	18.422	18.433	−14.57	99.76	85.19
−90	3.4229	0.01077	13.224	13.235	−12.22	98.38	86.16
−80	4.7822	0.01088	9.6840	9.6949	−9.85	96.98	87.13
−70	6.5522	0.01100	7.2208	7.2318	−7.49	95.59	88.10
−60	8.8180	0.01111	5.4733	5.4844	−5.01	94.08	89.07
−50	11.674	0.01124	4.2112	4.2224	−2.54	92.56	90.02
−40	15.222	0.01136	3.2843	3.2957	−0.03	91.01	90.97
−30	19.773	0.01150	2.5934	2.6049	2.51	89.40	91.91
−20	24.845	0.01163	2.0710	2.0826	5.13	87.82	92.95
−10	31.162	0.01178	1.6707	1.6825	7.68	86.07	93.75
0	38.657	0.01193	1.3604	1.3723	10.32	84.33	94.65
10	47.464	0.01209	1.1169	1.1290	13.00	82.53	95.53
20	57.727	0.01226	0.92405	0.93631	15.71	80.67	96.38
30	69.591	0.01243	0.76956	0.78208	18.45	78.76	97.21
40	83.206	0.01262	0.64491	0.65753	21.23	76.79	98.02
50	98.727	0.01281	0.54325	0.55606	24.04	74.75	98.79
60	116.31	0.01302	0.45970	0.47272	26.80	72.65	99.54
70	136.12	0.01325	0.39048	0.40373	29.78	70.46	100.24
80	158.33	0.01349	0.33272	0.34621	32.71	68.20	100.91
90	183.09	0.01375	0.28414	0.29789	35.69	65.84	101.53
100	210.60	0.01404	0.24298	0.25702	38.72	63.37	102.09
110	241.04	0.01435	0.20787	0.22222	41.81	60.78	102.59
120	274.60	0.01469	0.17768	0.19238	44.96	58.05	103.01
130	311.50	0.01508	0.15153	0.16661	48.19	55.14	103.33
140	351.94	0.01552	0.12866	0.14418	51.52	52.02	103.54
150	396.19	0.01602	0.10846	0.12448	54.97	48.63	103.60
160	444.53	0.01663	0.09038	0.10701	58.58	44.88	103.46
170	497.26	0.01737	0.07391	0.09128	62.42	40.62	103.04
180	554.78	0.01833	0.05846	0.07679	66.62	35.57	102.19
190	617.59	0.01973	0.04311	0.06284	71.46	29.09	100.55
200	686.36	0.02244	0.02500	0.04744	78.30	19.13	97.43
204.81	721.91	0.03053	0	0.03053	87.25	0	87.25

Table C.9a Saturated Refrigerant-22, Temperature Table (English Units) *continued*

Temp. °F T	Enthalpy (Btu/lbm)*			Entropy (Btu/(lbm · R))*		
	Sat. liq. h_f	Evap. h_{fg}	Sat. vapor h_g	Sat. liq. s_f	Evap. s_{fg}	Sat. vapor s_g
−150	−25.97	113.49	87.52	−0.07147	0.36648	0.29501
−140	−23.72	112.40	88.68	−0.06432	0.35161	0.28729
−130	−21.46	111.31	89.85	−0.05736	0.33763	0.28027
−120	−19.19	110.21	91.02	−0.05055	0.32443	0.27388
−110	−16.89	109.09	92.20	−0.04389	0.31194	0.26805
−100	−14.56	107.93	93.37	−0.03734	0.30008	0.26274
−90	−12.22	106.76	94.54	−0.03091	0.28878	0.25787
−80	−9.84	105.55	95.71	−0.02457	0.27799	0.25342
−70	−7.43	104.30	96.87	−0.01832	0.26764	0.24932
−60	−4.99	103.00	98.01	−0.01214	0.25770	0.24556
−50	−2.51	101.65	99.14	−0.00604	0.24813	0.24209
−40	0	100.26	100.26	0.00000	0.23888	0.23888
−30	2.55	98.80	101.35	0.00598	0.22993	0.23591
−20	5.13	97.29	102.42	0.01189	0.22126	0.23315
−10	7.75	95.70	103.45	0.01776	0.21282	0.23058
0	10.41	94.06	104.47	0.02357	0.20460	0.22817
10	13.10	92.34	105.44	0.02932	0.19660	0.22592
20	15.84	90.54	106.38	0.03503	0.18876	0.22379
30	18.61	88.67	107.28	0.04070	0.18108	0.22178
40	21.42	86.72	108.14	0.04632	0.17354	0.21986
50	24.27	84.68	108.95	0.05190	0.16613	0.21803
60	27.17	82.54	109.71	0.05745	0.15882	0.21627
70	30.12	80.29	110.41	0.06296	0.15160	0.21456
80	33.11	77.94	111.05	0.06846	0.14442	0.21288
90	36.16	75.46	111.62	0.07394	0.13728	0.21122
100	39.27	72.84	112.11	0.07942	0.13014	0.20956
110	42.45	70.05	112.50	0.08491	0.12296	0.20787
120	45.70	67.08	112.78	0.09042	0.11571	0.20613
130	49.06	63.88	112.94	0.09598	0.10833	0.20431
140	52.53	60.40	112.93	0.10163	0.10072	0.20235
150	56.14	56.59	112.73	0.10739	0.09281	0.20020
160	59.95	52.31	112.26	0.11334	0.08442	0.19776
170	64.02	47.40	111.42	0.11959	0.07531	0.19490
180	68.50	41.57	110.07	0.12635	0.06498	0.19133
190	73.71	34.02	107.73	0.13409	0.05237	0.18646
200	80.86	21.99	102.85	0.14460	0.03334	0.17794
204.81	91.33	0	91.33	0.16016	0	0.16016

* The enthalpy and entropy of saturated liquid R-22 are taken as zero at a temperature of −40°F.
Source: Reprinted by permission from Haberman, W. L., John, J. E. A., 1980. Engineering Thermodynamics. Allyn & Bacon, Boston, MA, pp. 414–415 (Table A.10).

Table C.9b Saturated Refrigerant-22, Temperature Table (Metric Units)

Temp. °C T	Abs. press. (kPa) p	Specific volume (m³/kg)			Internal energy (kJ/kg)		
		Sat. liq. v_f	Evap. v_{fg}	Sat. vapor v_g	Sat. liq. u_f	Evap. u_{fg}	Sat. vapor u_g
−100	2.0750	0.0006366	8.0083	8.0089	−59.37	246.86	187.49
−95	3.2323	0.0006418	5.2845	5.2851	−54.66	244.12	189.46
−90	4.8994	0.0006470	3.5804	3.5810	−49.92	241.36	191.36
−85	7.2412	0.0006525	2.4847	2.4854	−45.16	238.59	193.43
−80	10.461	0.0006581	1.7626	1.7633	−40.36	235.80	195.44
−75	14.794	0.0006638	1.2757	1.2764	−35.52	233.00	197.48
70	20.523	0.0006697	0.94033	0.94100	−30.62	230.13	199.51
−65	27.965	0.0006758	0.69876	0.70552	−25.68	227.21	201.53
−60	37.480	0.0006821	0.53649	0.53717	−20.68	224.25	203.57
−55	49.474	0.0006885	0.41416	0.41485	−15.62	221.22	205.60
−50	63.139	0.0006952	0.32387	0.32457	−10.50	218.52	208.02
−45	82.701	0.0007022	0.25630	0.25700	−5.32	214.94	209.62
−40	104.943	0.0007093	0.20505	0.20576	−0.07	211.68	211.61
−35	131.669	0.0007168	0.16569	0.16569	5.24	208.33	213.57
−30	163.470	0.0007245	0.13513	0.13585	10.60	204.91	215.51
−25	200.968	0.0007325	0.11113	0.11186	16.01	233.45	217.44
−20	244.814	0.0007409	0.092106	0.092847	21.55	197.77	219.32
−15	295.686	0.0007496	0.076878	0.077628	27.11	194.07	221.18
−10	354.284	0.0007587	0.064583	0.065342	32.74	190.25	222.99
−5	421.330	0.0007683	0.054573	0.055341	35.52	189.24	224.76
0	497.567	0.0007783	0.046359	0.047137	44.20	182.30	226.50
5	583.756	0.0007889	0.039568	0.040357	50.02	178.15	228.17
10	680.673	0.0008000	0.033915	0.034715	55.92	173.87	229.79
15	789.117	0.0008118	0.029177	0.029989	61.88	169.48	231.36
20	909.899	0.0008243	0.025180	0.026004	67.92	164.92	232.84
25	1043.856	0.0008376	0.021787	0.022625	74.04	160.22	234.26
30	1191.842	0.0008519	0.018891	0.019743	80.23	155.36	235.59
35	1354.741	0.0008673	0.016400	0.017267	86.52	150.30	236.82
40	1533.466	0.0008839	0.018859	0.015137	92.90	145.04	237.94
45	1728.969	0.0009020	0.012384	0.013286	99.40	139.53	238.93
50	1942.254	0.0009219	0.010751	0.011672	106.04	133.73	239.77
55	2174.382	0.0009440	0.009440	0.010257	112.81	127.62	240.43
60	2426.496	0.0009687	0.0080321	0.0090008	119.83	121.01	240.84
65	2699.843	0.0009970	0.0068907	0.0078877	127.04	113.93	240.97
70	2995.810	0.0010298	0.0058593	0.0068891	134.53	106.23	240.76
75	3316.03	0.0010691	0.0049144	0.0059835	142.43	97.62	240.05
80	3662.29	0.0011181	0.0040307	0.0051488	150.92	87.70	238.62
85	4036.81	0.0011832	0.0031751	0.0043583	160.31	75.79	236.10
90	4442.50	0.0012822	0.0022823	0.0035645	171.50	59.90	231.40
95	4883.49	0.001 5205	0.0010311	0.0025516	188.92	29.92	218.84
96.006	4977.39	0.0019056	0	0.0019056	203.09	0	203.09

* The enthalpy and entropy of saturated liquid R-22 are taken as zero at a temperature of −40°F.
Source: Reprinted by permission from Haberman, W. L., John, J. E. A., 1980. Engineering Thermodynamics. Allyn & Bacon, Boston, MA, pp. 472–473 (Table B.10).

Table C.9b Saturated Refrigerant-22, Temperature Table (Metric Units) *continued*

Temp. °C T	Enthalpy (kJ/kg)*			Entropy (kJ/(kg · K))*		
	Sat. liq. h_f	Evap. h_{fg}	Sat. vapor h_g	Sat. liq. s_f	Evap. s_{fg}	Sat. vapor s_g
−100	−59.37	263.48	204.11	−0.29317	1.52159	1.22842
−95	−54.66	261.20	206.54	−0.26426	1.46397	1.19971
−90	−49.92	258.91	208.98	−0.24016	1.41356	1.17340
−85	−45.16	256.59	211.43	−0.21447	1.36369	1.14922
−80	−40.35	254.25	213.89	−0.18928	1.31624	1.12696
−75	−35.51	251.87	216.36	−0.16452	1.27098	1.10646
−70	−30.61	249.42	218.82	−0.14012	1.22771	1.08759
−65	−25.66	246.92	221.26	−0.11611	1.18621	1.07010
−60	−20.65	244.35	223.70	−0.09234	1.14629	1.05395
−55	−15.59	241.70	226.12	−0.06891	1.10792	1.03901
−50	−10.46	238.96	228.51	−0.04569	1.07081	1.02512
−45	−5.26	236.13	230.87	−0.02276	1.03495	1.01219
−40	0.00	233.20	233.20	0.00000	1.00014	1.00014
−35	5.33	230.15	235.48	0.02251	0.96638	0.98889
−30	10.72	227.00	237.72	0.04485	0.93354	0.97839
−25	16.19	223.72	239.92	0.06699	0.90152	0.96851
−20	21.73	220.33	242.05	0.08895	0.87032	0.95927
−15	27.33	216.79	244.13	0.1 1075	0.83977	0.95052
−10	33.01	213.13	246.14	0.13234	0.80990	0.94224
−5	38.76	209.32	248.08	0.15380	0.78057	0.93437
0	44.59	205.36	249.95	0.17178	0.75178	0.92688
5	50.48	201.24	251.73	0.19627	0.72346	0.91973
10	56.46	196.96	253.42	0.21727	0.69559	0.91286
15	62.52	192.49	255.02	0.23819	0.66802	0.90621
20	68.67	187.84	256.50	0.25899	0.64074	0.89973
25	74.91	182.97	257.88	0.27970	0.61367	0.89337
30	81.25	177.87	259.12	0.30037	0.58676	0.88713
35	87.69	172.52	260.22	0.32104	0.55982	0.88086
40	94.26	166.89	261.15	0.34167	0.53291	0.87458
45	100.96	160.94	261.90	0.36233	0.50585	0.86818
50	107.83	154.62	262.44	0.38313	0.47844	0.86157
55	114.86	147.86	262.73	0.40409	0.45058	0.85467
60	122.18	140.50	262.68	0.42547	0.42171	0.84718
65	129.73	132.54	262.27	0.44714	0.39200	0.83914
70	137.62	123.77	261.40	0.46944	0.36071	0.83015
75	145.98	113.90	259.89	0.49267	0.32714	0.81981
80	155.01	102.47	257.48	0.51735	0.29016	0.80751
85	165.09	88.60	253.69	0.54446	0.24736	0.79182
90	177.04	70.04	247.24	0.57664	0.19288	0.76952
95	196.35	34.96	231.30	0.62731	0.09493	0.72224
96.006	212.57	0	212.57	0.67090	0	0.67090

* The enthalpy and the entropy of saturated liquid R-22 are taken as zero at a temperature of −40°C.
Source: Reprinted by permission from Haberman, W. L., John, J. E. A., 1980. Engineering Thermodynamics. Allyn & Bacon, Boston, MA, pp. 472–473 (Table B.10).

Table C.10a Superheated Refrigerant-22 Vapor (English Units)

Abs. press. (psia) P	Sat. temp. (°F) T	Sat. vapor	Temperature (°F)					
			−100	0	100	200	300	400
Specific volume, v (ft³/lbm)								
		v_g						
0.2	−155.80	188.29	223.01	285.14	347.22	409.32	471.37	533.42
0.5	−137.64	79.698	89.095	114.00	138.85	163.70	188.55	213.37
1	−122.16	41.678	44.458	56.949	69.397	81.83	94.27	106.68
2	−104.87	21.831	22.139	28.426	34.668	40.893	47.12	53.34
5	−78.62	9.3011	–	11.311	13.831	16.333	18.83	21.34
10	−55.58	4.8778	–	5.6060	6.8855	8.1464	9.399	10.65
15	−40.57	3.3412	–	3.7037	4.5701	5.4174	6.256	7.10
20	−29.12	2.5527	–	2.7521	3.4122	4.0529	4.685	5.325
40	1.63	1.3285	–	–	1.6749	2.0060	2.3281	2.657
60	22.03	0.90222	–	–	1.0952	1.3235	1.5424	1.767
80	37.76	0.68318	–	–	0.80477	0.98209	1.1495	1.323
100	50.77	0.54908	–	–	0.63003	0.77712	0.91372	1.046
120	61.95	0.45822	–	–	0.51309	0.64036	0.75651	0.8678
140	71.83	0.39243	–	–	0.42911	0.54258	0.64419	0.7409
160	80.71	0.34249	–	–	0.36568	0.46914	0.55993	0.6447
180	88.81	0.30323	–	–	0.31587	0.41194	0.49437	0.5715
200	96.27	0.27150	–	–	0.27553	0.36609	0.44190	0.51218
250	112.76	0.21351	–	–	–	0.28325	0.34740	0.40549
300	126.98	0.17400	–	–	–	0.22759	0.28431	0.33436
350	139.54	0.14514	–	–	–	0.18738	0.23917	0.28356
400	150.82	0.12297	–	–	–	0.15674	0.20524	0.24546
500	170.50	0.09053	–	–	–	0.11220	0.15757	0.19212
Internal energy, u (Btu/lbm)								
		u_g						
0.2	−155.80	79.88	85.37	96.49	109.20	123.42	139.07	156.01
0.5	−137.64	81.58	85.34	96.47	109.19	123.42	139.06	156.00
1	−122.16	83.05	85.30	96.45	109.18	123.40	139.04	155.98
2	−104.87	84.72	85.22	96.41	109.15	123.38	139.02	155.95
5	−78.62	87.27	–	96.30	109.07	123.33	138.96	155.87
10	−55.58	89.49	–	96.04	108.93	123.24	138.86	155.78
15	−40.57	90.92	–	95.81	108.80	123.15	138.80	155.69
20	−29.12	92.00	–	95.57	108.66	123.06	138.74	155.60
40	1.63	94.79	–	–	108.09	122.69	138.49	155.28
60	22.03	96.55	–	–	107.50	122.32	138.23	155.02
80	37.76	97.84	–	–	106.89	121.94	137.96	154.86
100	50.77	98.85	–	–	106.25	121.55	137.70	154.77
120	61.95	99.68	–	–	105.59	121.16	137.43	154.58
140	71.83	100.37	–	–	104.89	120.75	137.15	154.41
160	80.71	100.96	–	–	104.16	120.34	136.88	154.24
180	88.81	101.46	–	–	103.38	119.92	136.60	154.07
200	96.27	101.89	–	–	102.55	119.49	136.31	153.82
250	112.76	102.71	–	–	–	118.36	135.59	153.29
300	126.98	103.24	–	–	–	117.15	134.84	152.75
350	139.54	103.54	–	–	–	115.85	134.06	152.20
400	150.82	103.60	–	–	–	114.42	133.26	151.64
500	170.50	103.01	–	–	–	111.05	131.55	150.48

Table C.10a Superheated Refrigerant-22 Vapor (English Units)   *continued*

Abs. press. (psia) P	Sat. temp. (°F) T	Sat. vapor	Temperature (°F)					
			−100	0	100	200	300	400
Enthalpy, h (Btu/lbm)								
		hg						
0.2	−155.80	86.85	93.62	107.04	122.05	138.57	156.52	175.75
0.5	−137.64	88.96	93.59	107.02	122.04	138.56	156.51	175.74
1	−122.16	90.77	93.53	106.99	122.02	138.54	156.49	175.72
2	−104.87	92.80	93.42	106.93	121.98	138.51	156.46	175.69
5	−78.62	95.87	–	106.74	121.87	138.44	156.38	175.61
10	−55.58	98.52	–	106.41	121.67	138.31	156.26	175.49
15	−40.57	100.19	–	106.09	121.48	138.18	156.16	175.39
20	−29.12	101.44	–	105.76	121.28	138.06	156.08	175.31
40	1.63	104.63	–	–	120.49	137.54	155.72	174.95
60	22.03	106.57	–	–	119.66	137.01	155.35	174.64
80	37.76	107.95	–	–	118.80	136.48	154.98	174.45
100	50.77	109.01	–	–	117.91	135.93	154.61	174.13
120	61.95	109.85	–	–	116.98	135.38	154.23	173.85
140	71.83	110.54	–	–	116.01	134.81	153.84	173.60
160	80.71	111.10	–	–	114.99	134.23	153.46	173.33
180	88.81	111.56	–	–	113.90	133.64	153.06	173.10
200	96.27	111.93	–	–	112.75	133.03	152.67	172.78
250	112.76	112.59	–	–	–	131.46	151.66	172.05
300	126.98	112.90	–	–	–	129.78	150.62	171.31
350	139.54	112.94	–	–	–	127.98	149.55	170.51
400	150.82	112.70	–	–	–	126.02	148.45	169.81
500	170.50	111.38	–	–	–	121.43	146.13	168.26
Entropy, s (Btu/(lbm · R))								
		sg						
0.2	−155.80	0.29985	0.32028	0.35311	0.38261	0.40982	0.43515	0.45889
0.5	−137.64	0.28557	0.29917	0.33204	0.36155	0.38878	0.41411	0.43785
1	−122.16	0.27521	0.28314	0.31607	0.34561	0.37274	0.39807	0.42193
2	−104.87	0.26527	0.26700	0.30005	0.32964	0.35679	0.38212	0.40590
5	−78.62	0.25283	–	0.27872	0.30845	0.33566	0.36099	0.38477
10	−55.58	0.24399	–	0.26230	0.29229	0.31961	0.34494	0.36872
15	−40.57	0.23906	–	0.25248	0.28273	0.31016	0.33553	0.35931
20	−29.12	0.23566	–	0.24535	0.27588	0.30342	0.32884	0.35262
40	1.63	0.22780	–	–	0.25893	0.28694	0.31259	0.33637
60	22.03	0.22337	–	–	0.24855	0.27706	0.30293	0.32679
80	37.76	0.22029	–	–	0.24083	0.26987	0.29598	0.31998
100	50.77	0.21790	–	–	0.23454	0.26415	0.29050	0.31464
120	61.95	0.21593	–	–	0.22912	0.25936	0.28595	0.31024
140	71.83	0.21425	–	–	0.22428	0.25520	0.28205	0.30648
160	80.71	0.21276	–	–	0.21984	0.25149	0.27862	0.30320
180	88.81	0.21142	–	–	0.21566	0.24813	0.27554	0.30026
200	96.27	0.21018	–	–	0.21165	0.24503	0.27274	0.29760
250	112.76	0.20740	–	–	–	0.23814	0.26665	0.29186
300	126.98	0.20487	–	–	–	0.23204	0.26146	0.28705
350	139.54	0.20244	–	–	–	0.22641	0.25688	0.28286
400	150.82	0.20001	–	–	–	0.22104	0.25273	0.27914
500	170.50	0.19474	–	–	–	0.21034	0.24532	0.27267

Source: Reprinted by permission from Haberman, W. L., John, J. E. A., 1980. Engineering Thermodynamics. Allyn & Bacon, Boston, MA, pp. 416–417 (Table A. 11).

Table C.10b Superheated Refrigerant-22 Vapor (Metric Units)

Abs. press. (kPa)	Sat. temp. (°C) T_g	Sat. vapor	Temperature (°C)					
			−50	0	50	100	150	200
Specific volume, h (m³/kg)								
		v_g						
1	−107.61	15.885	21.455	26.263	31.070	35.876	40.685	45.492
5	−89.79	3.519	4.283	5.248	6.211	7.176	8.137	9.098
10	−80.64	1.8411	2.137	2.621	3.104	3.588	4.069	4.549
20	−70.41	0.9646	1.064	1.308	1.550	1.794	2.034	2.275
40	−58.86	0.5059	0.527	0.651	0.773	0.897	1.017	1.138
60	−51.37	0.3468	0.349	0.432	0.514	0.596	0.673	0.759
80	−45.68	0.2652	–	0.323	0.385	0.446	0.510	0.570
100	−41.03	0.2152	–	0.257	0.307	0.356	0.407	0.455
200	−25.12	0.1123	–	0.1260	0.1519	0.1770	0.2020	0.2260
300	−14.61	0.0766	–	0.0822	0.1001	0.1172	0.1331	0.1497
400	−6.52	0.0582	–	0.0598	0.0739	0.0871	0.0994	0.118
500	0.15	0.0469	–	–	0.0586	0.0694	0.0796	0.0895
600	5.88	0.0393	–	–	0.0482	0.0573	0.0660	0.0744
700	10.93	0.0338	–	–	0.0408	0.0488	0.0564	0.0637
800	15.47	0.0296	–	–	0.0347	0.0418	0.0489	0.0555
900	19.61	0.0263	–	–	0.0307	0.0373	0.0434	0.0493
1000	23.42	0.0237	–	–	0.0272	0.0333	0.0388	0.0442
1500	39.10	0.0155	–	–	0.0168	0.0213	0.0253	0.0294
2000	51.28	0.0113	–	–	–	0.0153	0.0184	0.0214
2500	61.38	0.0087	–	–	–	0.0116	0.0144	0.0169
3000	70.07	0.0069	–	–	–	0.0091	0.0117	0.0138
Internal energy, v (kJ/kg)								
		u_g						
1	−107.61	184.53	208.93	233.47	260.97	290.87	323.67	358.81
5	−89.79	191.51	208.88	233.44	260.95	290.86	323.66	358.80
10	−80.64	195.18	208.78	233.39	260.92	290.84	323.64	358.79
20	−70.41	199.32	208.56	233.24	260.82	290.82	323.63	358.77
40	−58.86	204.01	208.09	232.94	260.59	290.79	323.62	358.75
60	−51.37	207.04	207.64	232.66	260.40	290.73	323.58	358.72
80	−45.68	209.33	–	232.38	260.21	290.66	323.52	358.65
100	−41.03	211.19	–	232.11	260.02	290.52	323.43	358.56
200	−25.12	217.40	–	230.84	259.29	290.07	323.13	358.41
300	−14.61	221.31	–	229.38	258.36	289.42	322.77	358.11
400	−6.52	224.22	–	227.99	257.48	288.76	322.30	357.91
500	0.15	226.55	–	–	256.66	288.25	321.90	357.60
600	5.88	228.45	–	–	255.67	287.67	321.45	357.23
700	10.93	230.07	–	–	254.80	287.02	320.94	356.75
800	15.47	231.48	–	–	253.97	286.75	320.60	356:43
900	19.61	232.72	–	–	252.84	285.86	320.05	355.93
1000	23.42	233.75	–	–	251.61	285.04	319.61	355.51
1500	39.10	237.74	–	–	245.91	281.80	317.28	353.30
2000	51.28	239.94	–	–	–	278.10	315.07	352.20
2500	61.38	240.85	–	–	–	274.17	312.35	350.21
3000	70.07	240.68	–	–	–	269.41	309.39	348.33

Table C.10b Superheated Refrigerant-22 Vapor (Metric Units)   *continued*

Abs. press. (kPa)	Sat. temp. (°C)	Sat. vapor	Temperature (°C)					
	T_g		–50	0	50	100	150	200
Enthalpy, *h* (kJ/kg)								
		h_g						
1	–107.61	200.41	230.38	259.73	292.04	326.75	364.35	404.30
5	–89.79	209.10	230.29	259.68	292.00	326.74	364.34	404.29
10	–80.64	213.59	230.15	259.60	291.96	326.72	364.33	404.28
20	–70.41	218.61	229.84	259.40	291.82	326.70	364.32	404.27
40	–58.86	224.25	229.17	258.98	291.51	326.67	364.31	404.26
60	–51.37	227.85	228.56	258.59	291.25	326.51	364.30	404.25
80	–45.68	230.55	–	258.22	291.01	326.34	364.28	404.23
100	–41.03	232.71	–	257.81	290.81	326.12	364.08	404.05
200	–25.12	239.86	–	256.04	289.67	325.47	363.53	403.60
300	–14.61	244.29	–	254.04	288.39	324.58	362.72	403.02
400	–6.52	247.50	–	251.91	287.04	323.60	362.08	402.62
500	0.15	250.00	–	–	285.96	322.95	361.70	402.37
600	5.88	252.03	–	–	284.59	322.05	361.05	401.87
700	10.93	253.73	–	–	283.36	321.18	360.42	401.32
800	15.47	255.16	–	–	281.73	320.19	359.72	400.82
900	19.61	256.39	–	–	280.47	319.43	359.11	400.31
1000	23.42	257.45	–	–	278.81	318.34	358.41	399.66
1500	39.10	260.99	–	–	271.11	313.75	355.23	397.40
2000	51.28	262.54	–	–	–	308.70	351.87	395.00
2500	61.38	262.60	–	–	–	303.17	348.35	392.46
3000	70.07	261.38	–	–	–	296.71	344.49	389.73
Entropy, *s* (kJ/(kg · K))								
		s_g						
1	–107.61	1.27520	1.43157	1.55006	1.65851	1.75784	1.85237	1.94160
5	–89.79	1.17223	1.27655	1.39517	1.50370	1.60310	1.69763	1.78686
10	–80.64	1.12969	1.20937	1.32828	1.43689	1.53645	1.63098	1.72049
20	–70.41	1.08905	1.14178	1.26115	1.36998	1.46981	1.56434	1.65417
40	–58.86	1.05047	1.07252	1.19285	1.30205	1.40317	1.49770	1.58763
60	–51.37	1.02880	1.03197	1.15310	1.26268	1.36402	1.45855	1.54858
80	–45.68	1.01390	–	1.12447	1.23451	1.33601	1.43054	1.52067
100	–41.03	1.00240	–	1.10201	1.21251	1.31426	1.40924	1.49902
200	–25.12	0.96876	–	1.03090	1.14387	1.25603	1.35131	1.44089
300	–14.61	0.94985	–	0.98699	1.10230	1.20627	1.30160	1.39123
400	–6.52	0.93671	–	0.95345	1.07131	1.17637	1.27173	1.36163
500	0.15	0.92667	–	–	1.04749	1.15389	1.25125	1.34573
600	5.88	0.91851	–	–	1.02686	1.13460	1.23263	1.32711
700	10.93	0.91160	–	–	1.00967	1.11845	1.21698	1.31146
800	15.47	0.90558	–	–	0.99155	1.10218	1.20155	1.29603
900	19.61	0.90022	–	–	0.97826	1.09031	1.19035	1.28483
1000	23.42	0.89537	–	–	0.96455	1.07831	1.17906	1.27354
1500	39.10	0.87571	–	–	0.90752	1.03033	1.13463	1.22911
2000	51.28	0.85985	–	–	–	0.99257	1.10115	1.19750
2500	61.38	0.84625	–	–	–	0.96324	1.07357	1.17206
3000	70.07	0.83002	–	–	–	0.92890	1.04919	1.15023

Source: Reprinted by permission from Haberman, W. L., John, J. E. A., 1980. Engineering Thermodynamics. Allyn & Bacon, Boston, MA, pp. 474–475 (Table B.11).

Table C.11a Saturated Mercury, Pressure Table (English Units)

Sat. press., psia	Sat. temp., °F	Specific volume, ft³/lbm		Enthalpy, Btu/lbm			Entropy, Btu/(lbm · R)		
		v_f	v_g	h_f	h_{fg}	h_g	s_f	s_{fg}	s_g
0.01	233.57	1.21×10^{-3}	3637	6.668	127.732	134.400	0.01137	0.18428	0.19565
0.02	259.88	1.21	1893	7.532	127.614	135.146	0.01259	0.17735	0.18994
0.03	276.22	1.21	1292	8.068	127.540	135.608	0.01332	0.17332	0.18664
0.05	297.97	1.21	799	8.778	127.442	136.220	0.01427	0.16821	0.18248
0.1	329.73	1.22	416	9.814	127.300	137.114	0.01561	0.16126	0.17687
0.2	364.25	1.22×10^{-3}	217.3	10.936	127.144	138.080	0.01699	0.15432	0.17131
0.3	385.92	1.22	148.6	11.639	127.047	138.686	0.01783	0.15024	0.16807
0.4	401.98	1.22	113.7	12.159	126.975	139.134	0.01844	0.14736	0.16580
0.5	415.00	1.22	92.18	12.568	126.916	139.484	0.01892	0.14511	0.16403
0.6	425.82	1.23	77.84	12.929	126.868	139.797	0.01932	0.14328	0.16260
0.8	443.50	1.23×10^{-3}	59.58	13.500	126.788	140.288	0.01994	0.14038	0.16032
1	457.72	1.24	48.42	13.959	126.724	140.683	0.02045	0.13814	0.15859
2	504.93	1.24	25.39	15.476	126.512	141.988	0.02205	0.13116	0.15321
3	535.25	1.24	17.50	16.439	126.377	142.816	0.02302	0.12706	0.15008
5	575.7	1.24	10.90	17.741	126.193	143.934	0.02430	0.12188	0.14618
7	604.7	1.25×10^{-3}	8.04	18.657	126.065	144.722	0.02516	0.11846	0.14362
10	637.0	1.25	5.81	19.685	125.919	145.604	0.02610	0.11483	0.14093
20	706.0	1.26	3.09	21.864	125.609	147.473	0.02800	0.10779	0.13579
40	784.4	1.27	1.648	24.345	125.255	149.600	0.03004	0.10068	0.13072
60	835.7	1.28	1.144	25.940	125.024	150.964	0.03127	0.09652	0.12779
80	874.8	1.29×10^{-3}	0.885	27.149	124.849	152.008	0.03218	0.09356	0.12574
100	906.8	1.29	0.725	28.152	124.706	152.858	0.03290	0.09127	0.12417
150	969.4	1.30	0.507	30.090	124.424	154.514	0.03425	0.08707	0.12132
200	1017.2	1.31	0.392	31.560	124.209	155.769	0.03523	0.08411	0.11934
250	1057.2	1.31	0.322	32.784	124.029	156.813	0.03603	0.08178	0.11781
300	1091.2	1.32×10^{-3}	0.276	33.824	123.876	157.700	0.03669	0.07989	0.11658
400	1148.4	1.32	0.215	35.565	123.620	159.185	0.03775	0.07688	0.11463
450	1173.2	1.32	0.194	36.315	123.509	159.824	0.03820	0.07566	0.11386
500	1196.0	1.33	0.177	37.006	123.406	160.412	0.03861	0.07455	0.11316
600	1236.8	1.34	0.151	38.245	123.221	161.466	0.03932	0.07264	0.11196
700	1273.3	1.34×10^{-3}	0.132	39.339	123.058	162.397	0.03993	0.07102	0.11095
800	1306.1	1.34	0.118	40.324	122.910	163.234	0.04047	0.06961	0.11008
900	1336.2	1.35	0.106	41.226	122.775	164.001	0.04095	0.06837	0.10932
1000	1364.0	1.35	0.098	42.056	122.649	164.705	0.04139	0.06726	0.10865
1100	1390.0	1.36	0.090	42.828	122.533	165.361	0.04179	0.06625	0.10804

Sources: Reprinted by permission from Sheldon, L. A. Thermodynamic Properties of Mercury Vapor. General Electric Company, 1948. Liquid densities from WADC TR-59-598. Also from Reynolds, W. C., Perkins, H. C., 1977. Engineering Thermodynamics, second ed. McGraw-Hill, New York, p. 638 (Table B-3).

Table C.11b Saturated Mercury, Pressure Table (Metric Units)

Sat. press., MPa	Sat. temp., °C	Specific volume, m³/lbm v_g	Enthalpy, kJ/kg			Entropy, kJ/(kg · K)		
			h_f	h_{fg}	h_g	s_f	s_{fg}	s_g
0.000 06	109.2	259.6	15.13	297.20	312.33	0.0466	0.7774	0.8240
0.000 07	112.3	224.3	15.55	297.14	312.69	0.0477	0.7709	0.8186
0.000 08	115.0	197.7	15.93	297.09	313.02	0.0487	0.7654	0.8141
0.000 09	117.5	176.8	16.27	297.04	313.31	0.0496	0.7604	0.8100
0.000 10	119.7	160.1	16.58	297.00	313.58	0.0503	0.7560	0.8063
0.0002	134.9	83.18	18.67	296.71	315.38	0.0556	0.7271	0.7827
0.0004	151.5	43.29	20.93	296.40	317.33	0.0610	0.6981	0.7591
0.0006	161.8	29.57	22.33	296.21	318.54	0.0643	0.6811	0.7454
0.0008	169.4	22.57	23.37	296.06	319.43	0.0666	0.6690	0.7356
0.0010	175.5	18.31	24.21	295.95	320.16	0.0685	0.6596	0.7281
0.002	195.6	9.570	26.94	295.57	322.51	0.0744	0.6305	0.7049
0.004	217.7	5.013	29.92	295.15	325.07	0.0806	0.6013	0.6819
0.006	231.6	3.438	31.81	294.89	326.70	0.0843	0.5842	0.6685
0.008	242.0	2.632	33.21	294.70	327.91	0.0870	0.5721	0.6591
0.010	250.3	2.140	34.33	294.54	328.87	0.0892	0.5627	0.6519
0.02	278.1	1.128	38.05	294.02	332.07	0.0961	0.5334	0.6295
0.04	309.1	0.5942	42.21	293.43	335.64	0.1034	0.5039	0.6073
0.06	329.0	0.4113	44.85	293.06	337.91	0.1078	0.4869	0.5947
0.08	343.9	0.3163	46.84	292.78	339.62	0.1110	0.4745	0.5855
0.1	356.1	0.2581	48.45	292.55	341.00	0.1136	0.4649	0.5785
0.2	397.1	0.1377	53.87	291.77	345.64	0.1218	0.4353	0.5571
0.3	423.8	0.095 51	57.38	291.27	348.65	0.1268	0.4179	0.5447
0.4	444.1	0.073 78	60.03	290.89	350.92	0.1305	0.4056	0.5361
0.5	460.7	0.060 44	62.20	290.58	352.78	0.1334	0.3960	0.5294
0.6	474.9	0.051 37	64.06	290.31	354.37	0.1359	0.3881	0.5240
0.7	487.3	0.044 79	65.66	290.08	355.74	0.1380	0.3815	0.5195
0.8	498.4	0.039 78	67.11	289.87	356.98	0.1398	0.3757	0.5155
0.9	508.5	0.035 84	68.42	289.68	358.10	0.1415	0.3706	0.5121
1.0	517.8	0.032 66	69.61	289.50	359.11	0.1429	0.3660	0.5089
1.2	534.4	0.027 81	71.75	289.19	360.94	0.1455	0.3581	0.5036
1.4	549.0	0.024 29	73.63	288.92	362.55	0.1478	0.3514	0.4992
1.6	562.0	0.021 61	75.37	288.67	364.04	0.1498	0.3456	0.4954
1.8	574.0	0.019 49	76.83	288.45	365.28	0.1515	0.3405	0.4920
2.0	584.9	0.017 78	78.23	288.24	366.47	0.1531	0.3359	0.4890
2.2	595.1	0.016 37	79.54	288.05	367.59	0.1546	0.3318	0.4864
2.4	604.6	0.015 18	80.75	287.87	368.62	0.1559	0.3280	0.4839
2.6	613.5	0.014 16	81.89	287.70	369.59	0.1571	0.3245	0.4816
2.8	622.0	0.013 29	82.96	287.54	370.50	0.1583	0.3212	0.4795
3.0	630.0	0.012 52	83.97	287.39	371.36	0.1594	0.3182	0.4776
3.5	648.5	0.010 96	86.33	287.04	373.37	0.1619	0.3115	0.4734
4.0	665.1	0.009 78	88.43	286.73	375.16	0.1641	0.3056	0.4697
4.5	680.3	0.008 85	90.35	286.44	376.79	0.1660	0.3004	0.4664
5.0	694.4	0.008 09	92.11	286.18	378.29	0.1678	0.2958	0.4636
5.5	707.4	0.007 46	93.76	285.93	379.69	0.1694	0.2916	0.4610
6.0	719.7	0.006 93	95.30	285.70	381.00	0.1709	0.2878	0.4587
6.5	731.3	0.006 48	96.75	285.48	382.23	0.1723	0.2842	0.4565
7.0	742.3	0.006 09	98.12	285.28	383.40	0.1736	0.2809	0.4545
7.5	752.7	0.005 75	99.42	285.08	384.50	0.1748	0.2779	0.4527

Source: Reprinted by permission from Karlekar, B.V., 1983. Thermodynamics for Engineers. Prentice-Hall, Englewood Cliffs, NJ, pp. 554–555 (Appendix E). Also reprinted by permission from Sheldon, L. A. Thermodynamic Properties of Mercury Vapor, General Electric Company, 1948.

Table C.12a Critical Point Data (English Units)

Substance	Formula	Molecular mass	Temp. R	Pressure psia	Volume, ft³/lb mole
Air	–	28.97	238.3	547.6	1.481
Ammonia	NH_3	17.03	729.8	1636	1.16
Argon	Ar	39.948	272	705	1.20
Bromine	Br_2	159.808	1052	1500	2.17
Carbon dioxide	CO_2	44.01	547.5	1071	1.51
Carbon monoxide	CO	28.011	240	507	1.49
Chlorine	Cl_2	70.906	751	1120	1.99
Deuterium (normal)	D_2	4.00	69.1	241	–
Helium	He	4.003	9.5	33.2	0.926
Helium³	He	3.00	6.01	16.9	–
Hydrogen (normal)	H_2	2.016	59.9	188.1	1.04
Krypton	Kr	83.80	376.9	798	1.48
Neon	Ne	20.183	80.1	395	0.668
Nitrogen	N_2	28.013	227.1	492	1.44
Nitrous oxide	N_2O	44.013	557.4	1054	1.54
Oxygen	O_2	31.999	278.6	736	1.25
Sulfur dioxide	SO_2	64.063	775.2	1143	1.95
Water	H_2O	18.015	1165.3	3204	0.90
Xenon	Xe	131.30	521.55	852	1.90
Benzene	C_6H_6	78.115	1012	714	4.17
n-Butane	C_4H_{10}	58.124	765.2	551	4.08
Carbon tetrachloride	CCl_4	153.82	1001.5	661	4.42
Chloroform	$CHCl_3$	119.38	965.8	794	3.85
Dichlorodifluoromethane	CCl_2F_2	120.91	692.4	582	3.49
Dichlorofluoromethane	$CHCl_2F$	102.92	813.0	749	3.16
Ethane	C_2H_6	30.020	549.8	708	2.37
Ethyl alcohol	C_2H_5OH	46.07	929.0	926	2.68
Ethylene	C_2H_4	28.054	508.3	742	1.99
n-Hexane	C_6H_{14}	86.178	914.2	439	5.89
Methane	CH_4	16.043	343.9	673	1.59
Methyl alcohol	CH_3OH	32.042	923.7	1154	1.89
Methyl chloride	CH_3Cl	50.488	749.3	968	2.29
Propane	C_3H_8	44.097	665.9	617	3.20
Propene	C_3H_6	42.081	656.9	670	2.90
Propyne	C_3H_4	40.065	722	776	–
Trichlorofluoromethane	CCl_3F	137.37	848.1	635	3.97

Sources: Reprinted with permission from Kobe, K.A., Lynn Jr., R. E., 1953. The Critical Properties of Elements and Compounds, Chem. Rev. 52, 117–236. Copyright 1953 American Chemical Society. Also reprinted from Van Wylen, G. J., Sonntag, R. E., 1986. Fundamentals of Classical Thermodynamics, third ed. Wiley, New York, p. 684 (Table A.6E). Copyright © 1986 John Wiley & Sons. Reprinted by permission of John Wiley & Sons, Inc.

Table C.12b Critical Point Data (Metric Units)

Substance	Formula	Molecular mass	Temp. K	Pressure MPa	Volume m^3/kgmole
Air	–	28.97	132.41	3.774	.0923
Ammonia	NH_3	17.03	405.5	11.28	.0724
Argon	Ar	39.948	151	4.86	.0749
Bromine	Br_2	159.808	584	10.34	.1355
Carbon dioxide	CO_2	44.01	304.2	7.39	.0943
Carbon monoxide	CO	28.011	133	3.50	.0930
Chlorine	Cl_2	70.906	417	7.71	.1242
Deuterium (normal)	D_2	4.00	38.4	1.66	–
Helium	He	4.003	5.3	0.23	.0578
Helium[3]	He	3.00	3.3	0.12	–
Hydrogen (normal)	H_2	2.016	33.3	1.30	.0649
Krypton	Kr	83.80	209.4	5.50	.0924
Neon	Ne	20.183	44.5	2.73	.0417
Nitrogen	N_2	28.013	126.2	3.39	.0899
Nitrous oxide	N_2O	44.013	309.7	7.27	.0961
Oxygen	O_2	31.999	154.8	5.08	.0780
Sulfur dioxide	SO_2	64.063	430.7	7.88	.1217
Water	H_2O	18.015	647.3	22.09	.0568
Xenon	Xe	131.30	289.8	5.88	.1186
Benzene	C_6H_6	78.115	562	4.92	.2603
n-Butane	C_4H_{10}	58.124	425.2	3.80	.2547
Carbon tetrachloride	CCl_4	153.82	556.4	4.56	.2759
Chloroform	$CHCl_3$	119.38	536.6	5.47	.2403
Dichlorodifluoromethane	CCl_2F_2	120.91	384.7	4.01	.2179
Dichlorofluoromethane	$CHCl_2F$	102.92	451.7	5.17	.1973
Ethane	C_2H_6	30.070	305.5	4.88	.1480
Ethyl alcohol	C_2H_5OH	46.07	516	6.38	.1673
Ethylene	C_2H_4	28.054	282.4	5.12	.1242
n-Hexane	C_6H_{14}	86.178	507.9	3.03	.3677
Methane	CH_4	16.043	191.1	4.64	.0993
Methyl alcohol	CH_3OH	32.042	513.2	7.95	.1180
Methyl chloride	CH_3Cl	50.488	416.3	6.68	.1430
Propane	C_3H_8	44.097	370	4.26	.1998
Propene	C_3H_6	42.081	365	4.62	.1810
Propyne	C_3H_4	40.065	401	5.35	–
Trichlorofluoromethane	CCl_3F	137.37	471.2	4.38	.2478

Sources: Reprinted with permission from Kobe, K.A., Lynn Jr., R. E., 1953. The Critical Properties of Elements and Compounds, Chem. Rev. 52, 117–236. Copyright 1953 American Chemical Society. Also reprinted from Van Wylen, G. J., Sonntag, R. E., 1986. Fundamentals of Classical Thermodynamics, third ed. Wiley, New York, p. 685 (Table A.6SI). Copyright © 1986 John Wiley & Sons. Reprinted by permission of John Wiley & Sons, Inc.

Table C.13a Gas Constant Data (English Units)

Substance	M lbm/ lbmole	c_p Btu/ (lbm·R)	$\bar{c}_p$ Btu/ lbmole R)	c_p Btu/ (lbm·R)	$\bar{c}_v$ Btu/ (lbmole·R)	R ft·lbf/ (lbm·R)	R Btu/ (lbm·R)	$k = c_p/c_v$
Argon, Ar	39.94	0.125	4.99	0.075	3.00	38.69	0.0497	1.67
Helium, He	4.003	1.24	4.96	0.744	2.97	386.0	0.4961	1.67
Hydrogen, H_2	2.016	3.42	6.89	2.435	4.90	766.5	0.9850	1.40
Nitrogen, N_2	28.02	0.248	6.95	0.177	4.96	55.15	0.0709	1.40
Oxygen, O_2	32.00	0.219	7.01	0.157	5.02	48.29	0.0621	1.39
Carbon monoxide, CO	28.01	0.249	6.97	0.178	4.98	55.17	0.0709	1.40
Air	28.97	0.240	6.95	0.172	4.96	53.34	0.0685	1.40
Water vapor, H_2O	18.016	0.446	8.04	0.336	6.05	85.78	0.1102	1.33
Methane, CH_4	16.04	0.532	8.53	0.408	6.54	96.3	0.1238	1.30
Carbon dioxide, CO_2	44.01	0.202	8.89	0.157	6.90	35.1	0.0451	1.29
Sulfur dioxide, SO_2	64.07	0.154	9.87	0.123	7.88	24.1	0.0310	1.25
Acetylene, C_2H_2	26.04	0.409	10.65	0.333	8.66	59.3	0.0763	1.23
Ethylene, C_2H_4	28.05	0.410	11.50	0.339	9.51	55.1	0.0708	1.21
Ethane, C_2H_6	30.07	0.422	12.69	0.356	10.70	51.4	0.0660	1.19
Propane, C_3H_8	44.09	0.404	17.81	0.358	15.82	35.0	0.0450	1.13
Isobutane, C_4H_{10}	58.12	0.420	24.41	0.386	22.42	26.0	0.0342	1.09

Note: $R = \mathcal{R}/M =$ and $c_v = c_p - R$.
Source: Reprinted with permission from Reynolds, W. C., Perkins, H. C., 1977. Engineering Thermodynamics, second ed. McGraw-Hill, New York, p. 641 (Table B-6a).

Table C.13b Gas Constant Data (Metric Units)

Substance	M kg/ kgmole	c_p kJ/(kg·K)	$\bar{c}_p$ kJ/ (kgmole·K)	c_v kJ/(kg·K)	$\bar{c}_v$ kJ/ (kgmole·K)	R kJ/(kg·K)	$k\ c_p/c_v$
Argon, Ar	39.94	0.523	20.89	0.315	12.57	0.208	1.67
Helium, He	4.003	5.200	20.81	3.123	12.50	2.077	1.67
Hydrogen, H_2	2.016	14.32	28.86	10.19	20.55	4.124	1.40
Nitrogen, N_2	28.02	1.038	29.08	0.742	20.77	0.296	1.40
Oxygen, O_2	32.00	0.917	29.34	0.657	21.03	0.260	1.39
Carbon monoxide, CO	28.01	1.042	29.19	0.745	20.88	0.297	1.40
Air	28.97	1.004	29.09	0.718	20.78	0.286	1.40
Water vapor, H_2O	18.016	1.867	33.64	1.406	25.33	0.461	1.33
Carbon dioxide, CO_2	44.01	0.845	37.19	0.656	28.88	0.189	1.29
Sulfur dioxide, SO_2	64.07	0.644	41.26	0.514	32.94	0.130	1.25
Methane, CH_4	16.04	2.227	35.72	1.709	27.41	0.518	1.30
Propane, C_3H_8	44.09	1.691	74.56	1.502	66.25	0.189	1.13

Note: The R values in this table were determined from $R = \mathcal{R}/M$, and the c_v values from $c_v = c_p - R$. For purposes of internal consistency, more digits have been reported than justified by the experimental data. In calculations of entropy changes using the perfect gas equation of state, it is recommended that the value of k be computed to calculator accuracy.
Source: Reprinted with permission from Reynolds, W. C., Perkins, H. C., 1977. Engineering Thermodynamics, second ed. McGraw-Hill, New York, p. 642 (Table B-6b).

Table C.14a Constant Pressure, Specific Heat Ideal Gas Temperature Relations (English Units)

Gas		Range R	Max. error %
N_2	$\bar{c}_p = 9.3355 - 122.56\theta^{-1.5} + 256.38\theta^{-2} - 196.08\theta^{-3}$	540–6300	0.43
O_2	$\bar{c}_p = 8.9465 + 4.8044 \times 10^{-3}\theta^{1.5} - 42.679\theta^{-1.5} + 56.615\theta^{-2}$	540–6300	0.30
H_2	$\bar{c}_p = 13.505 - 167.96\theta^{-0.75} + 278.44\theta^{-1} - 134.01\theta^{-1.5}$	540–6300	0.60
CO	$\bar{c}_p = 16.526 - 0.16841\theta^{0.75} - 47.985\theta^{-0.5} + 42.246\theta^{-0.75}$	540–6300	0.42
OH	$\bar{c}_p = 19.490 - 14.185\theta^{0.25} + 4.1418\theta^{0.75} - 1.0196\theta$	540–6300	0.43
NO	$\bar{c}_p = 14.169 - 0.40861\theta^{0.5} - 16877\theta^{-0.5} + 17.899\theta^{-1.5}$	540–6300	0.34
H_2O	$\bar{c}_p = 34.190 - 43.868\theta^{0.25} + 19.778\theta^{0.5} - 0.88407\theta$	540–6300	0.43
CO_2	$\bar{c}_p = -0.89286 + 7.2967\theta^{0.5} - 0.98074\theta + 5.7835 \times 10^{-3}\theta^2$	540–6300	0.19
NO_2	$\bar{c}_p = 11.005 + 51.650\theta^{-0.5} - 86.916\theta^{-0.75} + 55.580\theta^{-2}$	540–6300	0.26
CH_4	$\bar{c}_p = -160.82 + 105.10\theta^{0.25} - 5.9452\theta^{0.75} + 77.408\theta^{-0.5}$	540–3600	0.15
C_2H_4	$\bar{c}_p = -22.800 + 29.433\theta^{0.5} - 8.5185\theta^{0.75} + 43.683\theta^{-3}$	540–3600	0.07
C_2H_6	$\bar{c}_p = 1.648 + 4.124\theta - 0.153\theta^2 + 1.74 \times 10^{-3}\theta^3$	540–2700	0.83
C_3H_8	$\bar{c}_p = -0.966 + 7.279\theta - 0.3755\theta^2 + 7.58 \times 10^{-3}\theta^3$	540–2700	0.40
C_4H_{10}	$\bar{c}_p = 0.945 + 8.873\theta - 0.438\theta^2 + 8.36 \times 10^{-3}\theta^3$	540–2700	0.54

Note: $\bar{c}_p = \frac{Btu}{lbmole \cdot R}$ $\theta = \frac{T(Rankine)}{180}$.
Sources: Reprinted by permission of the authors from Scott, T.C., Sonntag, R. E., 1971. University of Michigan, unpublished, except C_2H_6, C_3H_8, C_4H_{10} from Kobe, K.A., 1949. Pet. Refiner 28 (2), 113. Also reprinted from Van Wylen, G. J., Sonntag, R. E., 1986. Fundamentals of Classical Thermodynamics, third ed. Wiley, New York, p. 688 (Table A.9E). Copyright © 1986 John Wiley & Sons. Reprinted by permission of John Wiley & Sons, Inc.

Table C.14b Constant Pressure Specific Heat Ideal Gas Temperature Relations (Metric Units)

Gas		Range K	Max. error %
N_2	$\bar{c}_p = 39.060 - 512.79\theta^{-1.5} + 1072.7\theta^{-2} - 820.40\theta^{-3}$	300–3500	0.43
O_2	$\bar{c}_p = 37.432 + 0.020\,102\theta^{1.5} - 178.57\theta^{-1.5} + 236.88\theta^{-2}$	300–3500	0.30
H_2	$\bar{c}_p = 56.505 - 702.74\theta^{-0.75} + 1\,165.0\theta^{-1} - 560.70\theta^{-1.5}$	300–3500	0.60
CO	$\bar{c}_p = 69.145 - 0.704\,63\theta^{0.75} - 200.77\theta^{-0.5} + 176.76\theta^{-0.75}$	300–3500	0.42
OH	$\bar{c}_p = 81.546 - 59.350\theta^{0.25} + 17.329\theta^{0.75} - 4.2660\theta$	300–3500	0.43
NO	$\bar{c}_p = 59.283 - 1.7096\theta^{0.5} - 70.613\theta^{-0.5} + 74.889\theta^{-1.5}$	300–3500	0.34
H_2O	$\bar{c}_p = 143.05 - 183.54\theta^{0.25} + 82.751\theta^{0.5} - 3.6989\theta$	300–3500	0.43
CO_2	$\bar{c}_p = -3.7357 + 30.529\theta^{0.5} - 4.1034\theta + 0.024198\theta^2$	300–3500	0.19
NO_2	$\bar{c}_p = 46.045 + 216.10\theta^{-0.5} - 363.66\theta^{-0.75} + 232.550\theta^{-2}$	300–3500	0.26
CH_4	$\bar{c}_p = -672.87 + 439.74\theta^{0.25} - 24.875\theta^{0.75} + 323.88\theta^{-0.5}$	300–2000	0.15
C_2H_4	$\bar{c}_p = -95.395 + 123.15\theta^{0.5} - 35.641\theta^{0.75} + 182.77\theta^{-3}$	300–2000	0.07
C_2H_6	$\bar{c}_p = 6.895 + 17.26\theta - 0.6402\theta^2 + 0.00728\theta^3$	300–1500	0.83
C_3H_8	$\bar{c}_p = -4.042 + 30.46\theta - 1.571\theta^2 + 0.03171\theta^3$	300–1500	0.40
C_4H_{10}	$\bar{c}_p = 3.954 + 37.12\theta - 1.833\theta^2 + 0.034\,98\theta^3$	300–1500	0.54

Note: $\bar{c}_p = \frac{kJ}{kgmole \cdot K}$ $\theta = \frac{T(Kelvin)}{100}$.
Sources: Reprinted by permission of the authors from Scott, T.C., Sonntag, R. E., 1971. University of Michigan, unpublished, except C_2H_6, C_3H_8, C_4H_{10} from Kobe, K.A., 1949. Pet. Refiner 28 (2), 113. Also reprinted from Van Wylen, G. J., Sonntag, R. E., 1986. Fundamentals of Classical Thermodynamics, third ed. Wiley, New York, p. 688 (Table A.9SI). Copyright © 1986 John Wiley & Sons. Reprinted by permission of John Wiley & Sons, Inc.

Table C.15a Equation of State Constants (English Units)

Gas	van der Waal constants[†]		
	R, ft·lbf/(lbm·R)	a, ft^4·lbf/lbm^2	b, ft^3/lbm
Air	53.34	867	0.020
Ammonia	90.74	7809	0.035
Butane (n)	26.59	2199	0.032
Carbon dioxide	35.11	1012	0.016
Carbon monoxide	55.17	1004	0.023
Helium	386.04	1132	0.093
Hydrogen	766.53	32820	0.212
Methane	96.32	4763	0.043
Nitrogen	55.16	933	0.022
Oxygen	48.29	721	0.016
Propane	35.04	2585	0.033
Water vapor	85.78	9130	0.027

[†]The values of R, a, and b can be obtained from $R = \mathscr{R}/M$, $a = (27/64)(RT_c)^2/p_c$, and $b = RT_c/(8p_c)$, where M is the molecular mass and T_c and p_c are the temperature and pressure at the critical state.

Source: Reprinted from Van Wylen, G. J., Sonntag, R. E., 1978. Fundamentals of Classical Thermodynamics, SI Version, second ed. Wiley, New York. Copyright © 1978 John Wiley & Sons. Reprinted by permission of John Wiley & Sons, Inc.

Table C.15a Equation of State Constants (English Units)   *continued*

Gas	Beattie-Bridgeman constants				
	$a \times 10^3$, ft^3/lbm	A_o, ft^4·lbf/lbm^2	$b \times 10^3$, ft^3/lbm	$B_o \times 10^2$, ft^3/lbm	$c \times 10^{-6}$, (ft·R)^3/lbm
Air	10.7	842.7	−24.7	0.255	0.1398
Ammonia	160.0	4480.7	179.8	3.21	26.1656
Butane (n)	33.5	286.1	26.0	6.79	5.6266
Carbon dioxide	26.0	1404.3	26.3	3.81	1.4008
Carbon monoxide	14.9	930.6	−0.395	2.88	0.1399
Helium	239.0	733.2	0.0	5.60	0.0009
Hydrogen	−40.2	26338.5	−346.2	16.67	0.0234
Methane	18.5	4809.8	−15.8	5.58	0.7469
Nitrogen	14.9	929.9	−3.96	2.88	0.1398
Oxygen	12.8	790.7	2.11	0.232	0.1400
Propane	26.6	333.0	15.6	6.58	2.5430

Source: Reprinted from Van Wylen, G. J., Sonntag, R. E., 1978. Fundamentals of Classical Thermodynamics, SI Version, second ed. Wiley, New York. Copyright © 1978 John Wiley & Sons. Reprinted by permission of John Wiley & Sons, Inc.

Table C.15a Equation of State Constants (English Units)   *continued*

Gas	Redlich-Kwong constants[†]	
	a, (ft^3/lbm)^2(R)$^{1/2}$ (psia)	b, ft^3/lbm
Air	13520	0.01397
Ammonia	214970	0.02435
Butane (n)	61700	0.02222
Carbon dioxide	23970	0.01080
Carbon monoxide	15900	0.01571
Helium	3707	0.06646

Table C.15a Equation of State Constants (English Units)   *continued*

| Gas | Redlich-Kwong constants[†] | |
	a, (ft³/lbm)²(R)^{1/2} (psia)	*b*, ft³/lbm
Hydrogen	257500	0.14690
Methane	89750	0.02961
Nitrogen	14270	0.01532
Oxygen	12190	0.01100
Propane	67590	0.02275
Water vapor	316030	0.01877

[†]These can be calculated from critical state data as $a = (2.9686 \times 10^{-3})(R^2 T_c^{2.5}/p_c)$ and $b = (6.0167 \times 10^{-4})(RT_c/p_c)$, where R is the specific gas constant in ft·lbf/(lbm·R), T_c is in R, and p_c is in psia.
Source: Reprinted from Van Wylen, G. J., Sonntag, R. E., 1978. Fundamentals of Classical Thermodynamics, SI Version, second ed. Wiley, New York. Copyright © 1978 John Wiley & Sons. Reprinted by permission of John Wiley & Sons, Inc.

Table C.15b Equations of State Constants (SI Units)

| Gas | van der Waal constants[†] | | |
	R, kJ/(kg·K)	*a*, kN·m⁴/kg²	*b* × 10³, m³/kg
Air	0.287	0.162	1.25
Ammonia	0.488	1.457	2.18
Butane (*n*)	0.143	0.410	2.00
Carbon dioxide	0.189	0.189	1.00
Carbon monoxide	0.297	0.187	1.44
Helium	2.077	0.211	5.81
Hydrogen	4.124	6.125	13.24
Methane	0.518	0.889	2.68
Nitrogen	0.297	0.174	1.37
Oxygen	0.260	0.135	1.00
Propane	0.189	0.482	2.06
Water vapor	0.462	1.704	1.69

[†]The values of R, a, and b can be obtained from $R = \mathscr{R}/M$, $a = (27/64)(RT_c)^2/p_c$, and $b = RT_c/(8p_c)$, where M is the molecular mass and T_c and p_c are the temperature and pressure at the critical state.
Source: Reprinted from Van Wylen, G. J., Sonntag, R. E., 1978. Fundamentals of Classical Thermodynamics, SI Version, second ed. Wiley, New York. Copyright © 1978 John Wiley & Sons. Reprinted by permission of John Wiley & Sons, Inc.

Table C.15b Equations of State Constants (SI Units)   *continued*

| Gas | Beattie-Bridgeman constants | | | | |
	a × 10⁴, m³/kg	*A*₀, m⁴·kN/kg²	*b* × 10⁴, m³/kg	*B*₀, m³/kg	*c* × 10⁻⁶, (m·K)³/kg
Air	6.66	0.157	−15.3	0.159	14.98
Ammonia	100.1	0.836	112.0	0.200	2800.82
Butane (*n*)	20.92	0.053	16.2	0.424	602.29
Carbon dioxide	16.22	0.262	16.5	0.238	149.94
Carbon monoxide	9.35	0.174	−0.25	0.180	14.96
Helium	149.3	0.137	0.0	0.350	0.0997
Hydrogen	−25.1	4.914	−216.0	1.042	2.50
Methane	11.53	0.897	−9.85	0.349	79.93
Nitrogen	9.35	0.173	−2.47	0.180	14.96
Oxygen	8.00	0.147	1.31	0.145	15.00
Propane	16.60	0.062	9.73	0.411	272.22

Source: Reprinted from Van Wylen, G. J., Sonntag, R. E., 1978. Fundamentals of Classical Thermodynamics, SI Version, second ed. Wiley, New York. Copyright © 1978 John Wiley & Sons. Reprinted by permission of John Wiley & Sons, Inc.

Table C.15b Equations of State Constants (SI Units) *continued*

Gas	Redlich-Kwong constants†	
	a, (m³/kg)² (K)^(1/2)(kPa)	b × 10³, m³/kg
Air	1.881	0.8721
Ammonia	29.91	1.520
Butane (n)	8.584	1.387
Carbon dioxide	3.335	0.6742
Carbon monoxide	2.212	0.9808
Helium	0.5157	4.149
Hydrogen	35.82	9.169
Methane	12.49	1.848
Nitrogen	1.985	0.9564
Oxygen	1.696	0.6867
Propane	9.403	1.420
Water vapor	43.97	1.172

†These can be calculated from critical state data as $a = 0.42748(R^2 T_c^{2.5}/p_c)$ and $b = 0.08664(RT_c/p_c)$, where R is the specific gas constant in kJ/(kg·K), T_c is in K, and p_c is in kPa.
Source: Reprinted from Van Wylen, G. J., Sonntag, R. E., 1978. Fundamentals of Classical Thermodynamics, SI Version, second ed. Wiley, New York. Copyright © 1978 John Wiley & Sons. Reprinted by permission of John Wiley & Sons, Inc.

Table C.16a Air Tables (English Units)

Temp., R	h Btu/lbm	p_r	u Btu/lbm	v_r	ϕ Btu/(lbm·R)
200	47.67	0.04320	33.96	1714.9	0.36303
220	52.46	0.06026	37.38	1352.5	0.38584
240	57.25	0.08165	40.80	1088.8	0.40666
260	62.03	0.10797	44.21	892.0	0.42582
280	66.82	0.13986	47.63	741.6	0.44356
300	71.61	0.17795	51.04	624.5	0.46007
320	76.40	0.22290	54.46	531.8	0.47550
340	81.18	0.27545	57.87	457.2	0.49002
360	85.97	0.3363	61.29	396.6	0.50369
380	90.75	0.4061	64.70	346.6	0.51663
400	95.53	0.4858	68.11	305.0	0.52890
420	100.32	0.5760	71.52	270.1	0.54058
440	105.11	0.6776	74.93	240.6	0.55172
460	109.90	0.7913	78.36	215.33	0.56235
480	114.69	0.9182	81.77	193.65	0.57255
500	119.48	1.0590	85.20	174.90	0.58233
520	124.27	1.2147	88.62	158.58	0.59173
540	129.06	1.3860	92.04	144.32	0.60078
560	133.86	1.5742	95.47	131.78	0.60950
580	138.66	1.7800	98.90	120.70	0.61793
600	143.47	2.005	102.34	110.88	0.62607
620	148.28	2.249	105.78	102.12	0.63395
640	153.09	2.514	109.21	94.30	0.64159
660	157.92	2.801	112.67	87.27	0.64902
680	162.73	3.111	116.12	80.96	0.65621
700	167.56	3.446	119.58	75.25	0.66321
720	172.39	3.806	123.04	70.07	0.67002
740	177.23	4.193	126.51	65.38	0.67665
760	182.08	4.607	129.99	61.10	0.68312
780	186.94	5.051	133.47	57.20	0.68942
800	191.81	5.526	136.97	53.63	0.69558

Table C.16a Air Tables (English Units)   *continued*

Temp., R	h Btu/lbm	p_r	u Btu/lbm	v_r	ϕ Btu/(lbm · R)
820	196.69	6.033	140.47	50.35	0.70160
840	201.56	6.573	143.98	47.34	0.70747
860	206.46	7.149	147.50	44.57	0.71323
880	211.35	7.761	151.02	42.01	0.71886
900	216.26	8.411	154.57	39.64	0.72438
920	221.18	9.102	158.12	37.44	0.72979
940	226.11	9.834	161.68	35.41	0.73509
960	231.06	10.610	165.26	33.52	0.74030
980	236.02	11.430	168.83	31.76	0.74540
1000	240.98	12.298	172.43	30.12	0.75042
1020	245.97	13.215	176.04	28.59	0.75536
1040	250.95	14.182	179.66	27.17	0.76019
1060	255.96	15.203	183.29	25.82	0.76496
1080	260.97	16.278	186.93	24.58	0.76964
1100	265.99	17.413	190.58	23.40	0.77426
1120	271.03	18.604	194.25	22.30	0.77880
1140	276.08	19.858	197.94	21.27	0.78326
1160	281.14	21.18	201.63	20.293	0.78767
1180	286.21	22.56	205.33	19.377	0.79201
1200	291.30	24.01	209.05	18.514	0.79628
1220	296.41	25.53	212.78	17.700	0.80050
1240	301.52	27.13	216.53	16.932	0.80466
1260	306.65	28.80	220.28	16.205	0.80876
1280	311.79	30.55	224.05	15.518	0.81280
1300	316.94	32.39	227.83	14.868	0.81680
1320	322.11	34.31	231.63	14.253	0.82075
1340	327.29	36.31	235.43	13.670	0.82464
1360	332.48	38.41	239.25	13.118	0.82848
1380	337.68	40.59	243.08	12.593	0.83229
1400	342.90	42.88	246.93	12.095	0.83604
1420	348.14	45.26	250.79	11.622	0.83975
1440	353.37	47.75	254.66	11.172	0.84341
1460	358.63	50.34	258.54	10.743	0.84704
1480	363.89	53.04	262.44	10.336	0.85062
1500	369.17	55.86	266.34	9.948	0.85416
1520	374.47	58.78	270.26	9.578	0.85767
1540	379.77	61.83	274.20	9.226	0.86113
1560	385.08	65.00	278.13	8.890	0.86456
1580	390.40	68.30	282.09	8.569	0.86794
1600	395.74	71.73	286.06	8.263	0.87130
1620	401.09	75.29	290.04	7.971	0.87462
1640	406.45	78.99	294.03	7.691	0.87791
1660	411.82	82.83	298.02	7.424	0.88116
1680	417.20	86.82	302.04	7.168	0.88439
1700	422.59	90.95	306.06	6.924	0.88758
1720	428.00	95.24	310.09	6.690	0.89074
1740	433.41	99.69	314.13	6.465	0.89387
1760	438.83	104.30	318.18	6.251	0.89697
1780	444.26	109.08	322.24	6.045	0.90003
1800	449.71	114.03	326.32	5.847	0.90308
1820	455.17	119.16	330.40	5.658	0.90609
1840	460.63	124.47	334.50	5.476	0.90908

(Continued)

Table C.16a Air Tables (English Units)   *continued*

Temp., R	h Btu/lbm	p_r	u Btu/lbm	v_r	ϕ Btu/(lbm · R)
1860	466.12	129.95	338.61	5.302	0.91203
1880	471.60	135.64	342.73	5.134	0.91497
1900	477.09	141.51	346.85	4.974	0.91788
1920	482.60	147.59	350.98	4.819	0.92076
1940	488.12	153.87	355.12	4.670	0.92362
1960	493.64	160.37	359.28	4.527	0.92645
1980	499.17	167.07	363.43	4.390	0.92926
2000	504.71	174.00	367.61	4.258	0.93205
2020	510.26	181.16	371.79	4.130	0.93481
2040	515.82	188.54	375.98	4.008	0.93756
2060	521.39	196.16	380.18	3.890	0.94026
2080	526.97	204.02	384.39	3.777	0.94296
2100	532.55	212.1	388.60	3.667	0.94564
2120	538.15	220.5	392.83	3.561	0.94829
2140	543.74	229.1	397.05	3.460	0.95092
2160	549.35	238.0	401.29	3.362	0.95352
2180	554.97	247.2	405.53	3.267	0.95611
2200	560.59	256.6	409.78	3.176	0.95868
2220	566.23	266.3	414.05	3.088	0.96123
2240	571.86	276.3	418.31	3.003	0.96376
2260	577.51	286.6	422.59	2.921	0.96626
2280	583.16	297.2	426.87	2.841	0.96876
2300	588.82	308.1	431.16	2.765	0.97123
2320	594.49	319.4	435.46	2.691	0.97369
2340	600.16	330.9	439.76	2.619	0.97611
2360	605.84	342.8	444.07	2.550	0.97853
2380	611.53	355.0	448.38	2.483	0.98092
2400	617.22	367.6	452.70	2.419	0.98331

Notes: Reference level: 0 R and 1 atm.
Absolute entropies may be calculated from $s = \phi - 5 \ln(p/p°) + 1.0$ Btu/(lbm · R), where $p°$ is the reference pressure of 1 atm.
Sources: Abridged from Keenan, J. H., Kaye, J., 1948. Gas Tables. Wiley, New York. Copyright © 1948 John Wiley & Sons.
Reprinted by permission of John Wiley & Sons, Inc. Also reprinted by permission from Holman, J.P., 1980. Thermodynamics, third ed.
McGraw-Hill, New York, pp. 745–747 (Table A-17).

Table C.16b Air Tables (Metric Units)

T, K	h, kJ/kg	p_r	u, kJ/kg	v_r	ϕ kJ/(kg · K)
100	99.76	0.029 90	71.06	2230	1.4143
110	109.77	0.041 71	78.20	1758.4	1.5098
120	119.79	0.056 52	85.34	1415.7	1.5971
130	129.81	0.074 74	92.51	1159.8	1.6773
140	139.84	0.096 81	99.67	964.2	1.7515
150	149.86	0.123 18	106.81	812.0	1.8206
160	159.87	0.154 31	113.95	691.4	1.8853
170	169.89	0.190 68	121.11	594.5	1.9461
180	179.92	0.232 79	128.28	515.6	2.0033
190	189.94	0.281 14	135.40	450.6	2.0575
200	199.96	0.3363	142.56	396.6	2.1088
210	209.97	0.3987	149.70	351.2	2.1577
220	219.99	0.4690	156.84	312.8	2.2043
230	230.01	0.5477	163.98	280.0	2.2489
240	240.03	0.6355	171.15	251.8	2.2915

Table C.16b Air Tables (Metric Units)   *continued*

T, K	h, kJ/kg	p_r	u, kJ/kg	v_r	ϕ kJ/(kg · K)
250	250.05	0.7329	178.29	227.45	2.3325
260	260.09	0.8405	185.45	206.26	2.3717
270	270.12	0.9590	192.59	187.74	2.4096
280	280.14	1.0889	199.78	171.45	2.4461
290	290.17	1.2311	206.92	157.07	2.4813
300	300.19	1.3860	214.09	144.32	2.5153
310	310.24	1.5546	221.27	132.96	2.5483
320	320.29	1.7375	228.45	122.81	2.5802
330	330.34	1.9352	235.65	113.70	2.6111
340	340.43	2.149	242.86	105.51	2.6412
350	350.48	2.379	250.05	98.11	2.6704
360	360.58	2.626	257.23	91.40	2.6987
370	370.67	2.892	264.47	85.31	2.7264
380	380.77	3.176	271.72	79.77	2.7534
390	390.88	3.481	278.96	74.71	2.7796
400	400.98	3.806	286.19	70.07	2.8052
410	411.12	4.153	293.45	65.83	2.8302
420	421.26	4.522	300.73	61.93	2.8547
430	431.43	4.915	308.03	58.34	2.8786
440	441.61	5.332	315.34	55.02	2.9020
450	451.83	5.775	322.66	51.96	2.9249
460	462.01	6.245	329.99	49.11	2.9473
470	472.25	6.742	337.34	46.48	2.9693
480	482.48	7.268	344.74	44.04	2.9909
490	492.74	7.824	352.11	41.76	3.0120
500	503.02	8.411	359.53	39.64	3.0328
510	513.32	9.031	366.97	37.65	3.0532
520	523.63	9.684	374.39	35.80	3.0733
530	533.98	10.372	381.88	34.07	3.0930
540	544.35	11.097	389.40	32.45	3.1124
550	554.75	11.858	396.89	30.92	3.1314
560	565.17	11.659	404.44	29.50	3.1502
570	575.57	13.500	411.98	28.15	3.1686
580	586.04	14.382	419.56	26.89	3.1868
590	596.53	15.309	427.17	25.70	3.2047
600	607.02	16.278	434.80	24.58	3.2223
610	617.53	17.297	442.43	23.51	3.2397
620	628.07	18.360	450.13	22.52	3.2569
630	638.65	19.475	457.83	21.57	3.2738
640	649.21	20.64	465.55	20.674	3.2905
650	659.84	21.86	473.32	19.828	3.3069
660	670.47	23.13	481.06	19.026	3.3232
670	681.15	24.46	488.88	18.266	3.3392
680	691.82	25.85	496.65	17.543	3.3551
690	702.52	27.29	504.51	16.857	3.3707
700	713.27	28.80	512.37	16.205	3.3861
710	724.01	30.38	520.26	15.585	3.4014
720	734.20	31.92	527.72	15.027	3.4156
730	745.62	33.72	536.12	14.434	3.4314
740	756.44	35.50	544.05	13.900	3.4461
750	767.30	37.35	552.05	13.391	3.4607
760	778.21	39.27	560.08	12.905	3.4751
770	789.10	41.27	568.10	12.440	3.4894

(Continued)

Table C.16b Air Tables (Metric Units)   *continued*

T, K	h, kJ/kg	p_r	u, kJ/kg	v_r	ϕ kJ/(kg · K)
780	800.03	43.35	576.15	11.998	3.5035
790	810.98	45.51	584.22	11.575	3.5174
800	821.94	47.75	592.34	11.172	3.5312
810	832.96	50.08	600.46	10.785	3.5449
820	843.97	52.49	608.62	10.416	3.5584
830	855.01	55.00	616.79	10.062	3.5718
840	866.09	57.60	624.97	9.724	3.5850
850	877.16	60.29	633.21	9.400	3.5981
860	888.28	63.09	641.44	9.090	3.6111
870	899.42	65.98	649.70	8.792	3.6240
880	910.56	68.98	658.00	8.507	3.6367
890	921.75	72.08	666.31	8.233	3.6493
900	932.94	75.29	674.63	7.971	3.6619
910	944.15	78.61	682.98	7.718	3.6743
920	955.38	82.05	691.33	7.476	3.6865
930	966.64	85.60	699.73	7.244	3.6987
940	977.92	89.28	708.13	7.020	3.7108
950	989.22	93.08	716.57	6.805	3.7227
960	1000.53	97.00	725.01	6.599	3.7346
970	1011.88	101.06	733.48	6.400	3.7463
980	1023.25	105.24	741.99	6.209	3.7580
990	1034.63	109.57	750.48	6.025	3.7695
1000	1046.03	114.03	759.02	5.847	3.7810
1020	1068.89	123.12	775.67	5.521	3.8030
1040	1091.85	133.34	793.35	5.201	3.8259
1060	1114.85	143.91	810.61	4.911	3.8478
1080	1137.93	155.15	827.94	4.641	3.8694
1100	1161.07	167.07	845.34	4.390	3.8906
1120	1184.28	179.71	862.85	4.156	3.9116
1140	1207.54	193.07	880.37	3.937	3.9322
1160	1230.90	207.24	897.98	3.732	3.9525
1180	1254.34	222.2	915.68	3.541	3.9725
1200	1277.79	238.0	933.40	3.362	3.9922
1220	1301.33	254.7	951.19	3.194	4.0117
1240	1324.89	272.3	969.01	3.037	4.0308
1260	1348.55	290.8	986.92	2.889	4.0497
1280	1372.25	310.4	1004.88	2.750	4.0684
1300	1395.97	330.9	1022.88	2.619	4.0868
1320	1419.77	352.5	1040.93	2.497	4.1049
1340	1443.61	375.3	1059.03	2.381	4.1229
1360	1467.50	399.1	1077.17	2.272	4.1406
1380	1491.43	424.2	1095.36	2.169	4.1580
1400	1515.41	450.5	1113.62	2.072	4.1753
1420	1539.44	478.0	1131.90	1.9808	4.1923
1440	1563.49	506.9	1150.23	1.8942	4.2092
1460	1587.61	537.1	1168.61	1.8124	4.2258
1480	1611.80	568.8	1187.03	1.7350	4.2422
1500	1635.99	601.9	1205.47	1.6617	4.2585

Notes: Reference level: 0 K and 1 atm
Absolute entropies may be calculated from $s = \phi - 5\ln(p/p^o) + 4.1869$ kJ/(kg · K) where p^o is the reference pressure of 1 atm.
Sources: Adapted to SI units from Keenan, J. H., Kaye, J., 1948. Gas Tables. Wiley, New York. Copyright © 1948 John Wiley & Sons.
Reprinted by permission of John Wiley & Sons, Inc. Also reprinted by permission from Holman, J.P., 1980. Thermodynamics, third ed.
McGraw-Hill, New York, pp. 748–750 (Table A-17M).

Table C.16c Other gases (English Units)

Temp., R	Products of Combustion, 400% Theoretical Air		Products of Combustion, 200% Theoretical Air		Nitrogen		Oxygen		Water Vapor		Carbon Dioxide		Hydrogen		Carbon Monoxide	
	$\bar{h}$	$\bar{\phi}$	$\bar{h}$	$\bar{\phi}$	$\bar{h}$	$\bar{\phi}$	$\bar{h}$	$\bar{\phi}$	$\bar{h}$	$\bar{\phi}$	$\bar{h}$	$\bar{\phi}$	$\bar{h}$	$\bar{\phi}$	$\bar{h}$	$\bar{\phi}$
537	3746.8	46.318	3774.9	46.300	3729.5	45.755	3725.1	48.986	4258.3	45.079	4030.2	51.032	3640.3	31.194	3729.5	47.272
600	4191.9	47.101	4226.3	47.094	4167.9	46.514	4168.3	49.762	4764.7	45.970	4600.9	52.038	4075.6	31.959	3168.0	48.044
700	4901.7	48.195	4947.7	48.207	4864.9	47.588	4879.3	50.858	5575.4	47.219	5552.0	53.503	4770.2	33.031	4866.0	49.120
800	5617.5	49.150	5676.3	49.179	5564.4	48.522	5602.0	51.821	6396.9	48.316	6552.9	54.839	5467.1	33.961	5568.2	50.058
900	6340.3	50.002	6413.0	50.047	6268.1	49.352	6337.9	52.688	7230.9	49.298	7597.6	56.070	6165.3	34.784	6276.4	50.892
1000	7072.1	50.773	7159.8	50.833	6977.9	50.099	7087.5	53.477	8078.9	50.191	8682.1	57.212	6864.5	35.520	6992.2	51.646
1100	7812.9	51.479	7916.4	51.555	7695.0	50.783	7850.4	54.204	8942.0	51.013	9802.6	58.281	7564.6	36.188	7716.8	52.337
1200	8563.4	52.132	8683.6	52.222	8420.0	51.413	8625.8	54.879	9820.4	51.777	10955.3	59.283	8265.8	36.798	8450.8	52.976
1300	9324.1	52.741	9461.7	52.845	9153.9	52.001	9412.9	55.508	10714.5	52.494	12136.9	60.229	8968.7	37.360	9194.6	53.571
1400	10095.0	53.312	10250.7	53.430	9896.9	52.551	10210.4	56.099	11624.8	53.168	13344.7	61.124	9673.8	37.883	9948.1	54.129
1500	10875.6	53.851	11050.2	53.981	10648.9	53.071	11017.1	56.656	12551.4	53.808	14576.0	61.974	10381.5	38.372	10711.1	54.655
1600	11665.6	54.360	11859.6	54.504	11409.7	53.561	11832.5	57.182	13494.9	54.418	15829.0	62.783	11092.5	38.830	11483.4	55.154
1700	12464.3	54.844	12678.6	55.000	12178.9	54.028	12655.6	57.680	14455.4	54.999	17101.4	63.555	11807.4	39.264	12264.3	55.628
1800	13271.7	55.306	13507.0	55.473	12956.3	54.472	13485.8	58.155	15433.0	55.559	18391.5	64.292	12526.8	39.675	13053.2	56.078
1900	14087.2	55.747	14344.1	55.926	13741.6	54.896	14322.1	58.607	16427.5	56.097	19697.8	64.999	13250.9	40.067	13849.8	56.509
2000	14910.3	56.169	15189.3	56.360	14534.4	55.303	15164.0	59.039	17439.0	56.617	21018.7	65.676	13980.1	40.441	14653.2	56.922
2100	15740.5	56.574	16042.4	56.777	15334.0	55.694	16010.9	59.451	18466.9	57.119	22352.7	66.327	14714.5	40.799	15463.3	57.317
2200	16577.1	56.964	16902.5	57.177	16139.8	56.068	16862.6	59.848	19510.8	57.605	23699.0	66.953	15454.4	41.143	16279.4	57.696
2300	17419.8	57.338	17769.3	57.562	16951.2	56.429	17718.8	60.228	20570.6	58.077	25056.3	67.557	16199.8	41.475	17101.0	58.062
2400	18268.0	57.699	18642.1	57.933	17767.9	56.777	18579.2	60.594	21645.7	58.535	26424.0	68.139	16950.6	41.794	17927.4	58.414
2500	19121.4	58.048	19520.7	58.292	18589.5	57.112	19443.4	60.946	22735.4	58.980	27801.2	68.702	17707.3	42.104	18758.8	58.754
2600	19979.7	58.384	20404.6	58.639	19415.8	57.436	20311.4	61.287	23839.5	59.414	29187.1	69.245	18469.7	42.403	19594.3	59.081
2700	20842.8	58.710	21293.8	58.974	20246.4	57.750	21182.9	61.616	24957.2	59.837	30581.2	69.771	19237.8	42.692	20434.0	59.398
2800	21709.8	59.026	22187.5	59.300	21081.1	58.053	22057.8	61.934	26088.0	60.248	31982.8	70.282	20011.8	42.973	21277.2	59.705
2900	22581.4	59.331	23086.0	59.615	21919.5	58.348	22936.1	62.242	27231.2	60.650	33391.5	70.776	20791.5	43.247	22123.8	60.002
3000	23456.6	59.628	23988.5	59.921	22761.5	58.632	23817.1	62.540	28386.3	61.043	34806.6	71.255	21576.9	43.514	22973.4	60.290
3100	24335.5	59.916	24895.3	60.218	23606.8	58.910	24702.5	62.831	29552.8	61.426	36227.9	71.722	22367.7	43.773	23826.0	60.569

(Continued)

Table C.16c Other gases (English Units)   continued

Temp., R	Products of Combustion, 400% Theoretical Air		Products of Combustion, 200% Theoretical Air		Nitrogen		Oxygen		Water Vapor		Carbon Dioxide		Hydrogen		Carbon Monoxide	
	$\bar{h}$	$\bar{\phi}$	$\bar{h}$	$\bar{\phi}$	$\bar{h}$	$\bar{\phi}$	$\bar{h}$	$\bar{\phi}$	$\bar{h}$	$\bar{\phi}$	$\bar{h}$	$\bar{\phi}$	$\bar{h}$	$\bar{\phi}$	$\bar{h}$	$\bar{\phi}$
3200	25217.8	60.196	25805.6	60.507	24455.0	59.179	25590.5	63.113	30730.2	61.801	37654.7	72.175	23164.1	44.026	24681.2	60.841
3300	26102.9	60.469	26719.2	60.789	25306.0	59.442	26481.6	63.386	31918.2	62.167	39086.7	72.616	23965.5	44.273	25539.0	61.105
3400	26991.4	60.734	27636.4	61.063	26159.7	59.697	27375.9	63.654	33116.0	62.526	40523.6	73.045	24771.9	44.513	26399.3	61.362
3500	—	—	28556.8	61.329	27015.9	59.944	28273.3	63.914	34323.4	62.876	41965.2	73.462	25582.9	44.748	27261.8	61.612
3600	—	—	29479.9	61.590	27874.4	60.186	29173.9	64.168	35540.1	63.221	43411.0	73.870	26398.5	44.978	28126.6	61.855
3700	—	—	30406.0	61.843	28735.1	60.422	30077.5	64.415	36765.4	63.557	44860.6	74.267	27218.5	45.203	28993.5	62.093
3800	—	—	31334.8	62.091	29597.9	60.652	30984.1	64.657	37998.9	63.887	46314.0	74.655	28042.8	45.423	29862.3	62.325
3900	—	—	32266.2	62.333	30462.8	60.877	31893.6	64.893	39240.2	64.210	47771.0	75.033	28871.1	45.638	30732.9	62.551
4000	—	—	—	—	31329.4	61.097	32806.1	65.123	40489.1	64.528	49231.4	75.404	29703.5	45.849	31605.2	62.772
4100	—	—	—	—	32198.0	61.310	33721.6	65.350	41745.4	64.839	50695.1	75.765	30539.8	46.056	32479.1	62.988
4200	—	—	—	—	33068.1	61.520	34639.9	65.571	43008.4	65.144	52162.0	76.119	31379.8	46.257	33354.4	63.198
4300	—	—	—	—	33939.9	61.726	35561.1	65.788	44278.0	65.444	53632.1	76.464	32223.5	46.456	34231.2	63.405
4400	—	—	—	—	34813.1	61.927	36485.0	66.000	45553.9	65.738	55105.1	76.803	33070.9	46.651	35109.2	63.607
4500	—	—	—	—	35687.8	62.123	37411.8	66.208	46835.9	66.028	56581.0	77.135	33921.6	46.842	35988.6	63.805
4600	—	—	—	—	36563.8	62.316	38341.4	66.413	48123.6	66.312	58059.7	77.460	34775.7	47.030	36869.3	63.998
4700	—	—	—	—	37441.1	62.504	39273.6	66.613	49416.9	66.591	59541.1	77.779	35633.0	47.215	37751.0	64.188
4800	—	—	—	—	38319.5	62.689	40208.6	66.809	50715.5	66.866	61024.9	78.091	36493.4	47.396	38633.9	64.374
4900	—	—	—	—	39199.1	62.870	41146.1	67.003	52019.0	67.135	62511.3	78.398	37356.9	47.574	39517.8	64.556
5000	—	—	—	—	40079.8	63.049	42086.3	67.193	53327.4	67.401	64000.0	78.698	38223.8	47.749	40402.7	64.735
5100	—	—	—	—	40961.6	63.223	43029.1	67.380	54640.3	67.662	65490.9	78.994	39092.8	47.921	41288.6	64.910
5200	—	—	—	—	41844.4	63.395	43974.3	67.562	55957.4	67.918	66984.0	79.284	39965.1	48.090	42175.2	65.082
5300	—	—	—	—	42728.3	63.563	44922.2	67.743	57278.7	68.172	68479.1	79.569	40840.2	48.257	43063.2	65.252

Notes: To convert this table to metric units use the conversion factors 1 Btu/lbmole = 2.3258 kJ/kgmole, 1 Btu/(lbmole · R) = 4.1865 kJ/(kgmole K), and 1 R = $\frac{5}{9}$ K.

Reference level: 0 R and 1 atm; h in Btu/lbm mol; ϕ in Btu/(lbmole · R).

Absolute entropies at other pressures may be calculated from $\bar{s} = \bar{\phi} - 5 \ln(p/p°)$ Btu/(lbmole · R), where p° is the reference pressure of 1 atm.

Sources: Abridged from Keenan, J. H., Kaye, J., 1948. Gas Tables. Wiley, New York. Copyright ©1948 John Wiley & Sons. Reprinted by permission of John Wiley & Sons, Inc. Also reprinted by permission from Holman, J.P., 1980. Thermodynamics, third ed. McGraw-Hill, New York, pp. 751–752 (Table A-18).

Table C.17 Base-10 Logarithms of the Equilibrium Constants K_e for the Reaction

$$v_1 C_1 + v_2 C_2 \rightleftarrows v_3 C_3 + v_4 C_4 \quad K_e = \frac{\chi_3^{v_3} \chi_4^{v_4}}{\chi_1^{v_1} \chi_2^{v_2}} \left(\frac{p}{p_o}\right)^{v_3+v_4-v_1-v_2} \quad (p_o = 1 \text{ atm})$$

T, K	$H_2 \rightleftarrows 2H$	$O_2 \rightleftarrows 2O$	$H_2O \rightleftarrows H_2 + \frac{1}{2}O_2$	$H_2O \rightleftarrows OH + \frac{1}{2}H_2$	$CO_2 \rightleftarrows CO + \frac{1}{2}O_2$	$N_2 \rightleftarrows 2N$	$\frac{1}{2}O_2 + \frac{1}{2}N_2 \rightleftarrows NO$	$Na \rightleftarrows Na^+ + e^-$	$Cs \rightleftarrows Cs + e^-$
298	−71.210	−80.620	−40.047	−46.593	−45.043	−119.434	−15.187	−32.3	−25.1
400	−51.742	−58.513	−29.241	−33.910	−32.41	−87.473	−11.156	−24.3	−17.5
600	−32.667	−36.859	−18.663	−21.470	−20.07	−56.206	−7.219	−14.6	−10.0
800	−23.074	−25.985	−13.288	−15.214	−13.90	−40.521	−5.250	−9.58	−6.15
1000	−17.288	−19.440	−10.060	−11.444	−10.199	−31.084	−4.068	−6.54	−3.79
1200	−13.410	−15.062	−7.896	−8.922	−7.742	−24.619	−3.279	−4.47	−2.18
1400	−10.627	−11.932	−6.334	−7.116	−5.992	−20.262	−2.717	−2.97	−1.010
1600	−8.530	−9.575	−5.175	−5.758	−4.684	−16.869	−2.294	−1.819	−0.108
1800	−6.893	−7.740	−4.263	−4.700	−3.672	−14.225	−1.966	−0.913	+0.609
2000	−5.579	−6.269	−3.531	−3.852	−2.863	−12.016	−1.703	−0.175	+1.194
2200	−4.500	−5.064	−2.931	−3.158	−2.206	−10.370	−1.488	+0.438	+1.682
2400	−3.598	−4.055	−2.429	−2.578	−1.662	−8.992	−1.309	+0.956	+2.098
2600	−2.833	−3.206	−2.003	−2.087	−1.203	−7.694	−1.157	+1.404	+2.46
2800	−2.176	−2.475	−1.638	−1.670	−0.807	−6.640	−1.028	+1.792	+2.77
3000	−1.604	−1.840	−1.322	−1.302	−0.469	−5.726	−0.915	+2.13	+3.05
3200	−1.104	−1.285	−1.046	−0.983	−0.175	−4.925	−0.817	+2.44	+3.29
3500	−0.458	−0.571	−0.693	−0.557	+0.201	−3.893	−0.692	+2.84	+3.62
4000	+0.406	+0.382	−0.221	−0.035	+0.699	−2.514	−0.526	+3.38	+4.07
4500	+1.078	+1.125	+0.153	+0.392	+1.081	−1.437	−0.345	+3.82	+4.43
5000	+1.619	+1.719	+0.450	+0.799	+1.387	−0.570	−0.298	+4.18	+4.73

Source: Reprinted by permission from Reynolds, W. C., Perkins, H. C., 1977. Engineering Thermodynamics, second ed. McGraw-Hill, New York. p. 652 (Table B-14).

Table C.18 Isentropic Compressible Flow Tables for Air ($k = 1.4$)

M	p/p_{os}	T/T_{os}	ρ/ρ_{os}	A/A^*	M^*
0.10000	0.99303	0.99800	0.99502	5.82183	0.10944
0.20000	0.97250	0.99206	0.98028	2.96352	0.21822
0.30000	0.93947	0.98232	0.95638	2.03506	0.32572
0.40000	0.89561	0.96899	0.92427	1.59014	0.43133
0.50000	0.84302	0.95238	0.88517	1.33984	0.53452
0.60000	0.78400	0.93284	0.84045	1.18820	0.63481
0.70000	0.72093	0.91075	0.79158	1.09437	0.73179
0.80000	0.65602	0.88652	0.73999	1.03823	0.82514
0.90000	0.59126	0.86059	0.68704	1.00886	0.91460
1.00000	0.52828	0.83333	0.63394	1.00000	1.00000
1.10000	0.46835	0.80515	0.58170	1.00793	1.08124
1.20000	0.41238	0.77640	0.53114	1.03044	1.15828
1.30000	0.36091	0.74738	0.48290	1.06630	1.23114
1.40000	0.31424	0.71839	0.43742	1.11493	1.29987
1.50000	0.27240	0.68966	0.39498	1.17617	1.36458
1.60000	0.23527	0.66138	0.35573	1.25023	1.42539
1.70000	0.20259	0.63371	0.31969	1.33761	1.48247
1.80000	0.17404	0.60680	0.28682	1.43898	1.53598
1.90000	0.14924	0.58072	0.25699	1.55526	1.58609
2.00000	0.12780	0.55556	0.23005	1.68750	1.63299
2.10000	0.10935	0.53135	0.20580	1.83694	1.67687
2.20000	0.09352	0.50813	0.18405	2.00497	1.71791
2.30000	0.07997	0.48591	0.16458	2.19313	1.75629
2.40000	0.06840	0.46468	0.14720	2.40310	1.79218
2.50000	0.05853	0.44444	0.13169	2.63672	1.82574
2.60000	0.05012	0.42517	0.11787	2.89597	1.85714
2.70000	0.04295	0.40683	0.10557	3.18301	1.88653
2.80000	0.03685	0.38941	0.09463	3.50012	1.91404
2.90000	0.03165	0.37286	0.08489	3.84976	1.93981
3.00000	0.02722	0.35714	0.07623	4.23456	1.96396
3.50000	0.01311	0.28986	0.04523	6.78962	2.06419
4.00000	0.00659	0.23810	0.02766	10.71875	2.13809
4.50000	0.00346	0.19802	0.01745	16.56220	2.19360
5.00000	0.00189	0.16667	0.01134	25.00000	2.23607
5.50000	0.00107	0.14184	0.00758	36.86897	2.26913
6.00000	0.00063	0.12195	0.00519	53.17981	2.29528
6.50000	0.00039	0.10582	0.00364	75.13433	2.31626
7.00000	0.00024	0.09259	0.00261	104.14290	2.33333
7.50000	0.00016	0.08163	0.00190	141.84140	2.34738
8.00000	0.00010	0.07246	0.00141	190.10930	2.35907
8.50000	0.00007	0.06472	0.00107	251.08620	2.36889
9.00000	0.00005	0.05814	0.00082	327.18930	2.37722
9.50000	0.00003	0.05249	0.00063	421.13160	2.38433
10.00000	0.00002	0.04762	0.00049	535.93780	2.39046

Table C.19 Normal Shock Tables for Air ($k = 1.4$)

M_x	M_y	p_y/p_x	T_y/T_x	ρ_y/ρ_x	p_{osy}/p_{osx}	p_{osy}/p_x
0.00000	1.00000	1.00000	1.00000	1.00000	1.00000	1.89293
0.10000	0.91177	1.24500	1.06494	1.16908	0.99893	2.13285
0.20000	0.84217	1.51333	1.12799	1.34162	0.99280	2.40750
0.30000	0.78596	1.80500	1.19087	1.51570	0.97937	2.71359
0.40000	0.73971	2.12000	1.25469	1.68966	0.95820	3.04924
0.50000	0.70109	2.45833	1.32022	1.86207	0.92979	3.41327
0.60000	0.66844	2.82000	1.38797	2.03175	0.89520	3.80497
0.70000	0.64054	3.20500	1.45833	2.19772	0.85572	4.22383
0.80000	0.61650	3.61333	1.53158	2.35922	0.81268	4.66951
0.90000	0.59562	4.04500	1.60792	2.51568	0.76736	5.14178
1.00000	0.57735	4.50000	1.68750	2.66667	0.72087	5.64044
2.10000	0.56128	4.97833	1.77045	2.81190	0.67420	6.16538
2.20000	0.54706	5.48000	1.85686	2.95122	0.62814	6.71648
2.30000	0.53441	6.00500	1.94680	3.08455	0.58329	7.29368
2.40000	0.52312	6.55333	2.04033	3.21190	0.54014	7.89691
2.50000	0.51299	7.12500	2.13750	3.33333	0.49901	8.52613
2.60000	0.50387	7.72000	2.23834	3.44898	0.46012	9.18130
2.70000	0.49563	8.33833	2.34289	3.55899	0.42359	9.86240
2.80000	0.48817	8.98000	2.45117	3.66355	0.38946	10.56938
2.90000	0.48138	9.64500	2.56321	3.76286	0.35773	11.30224
3.00000	0.47519	10.33333	2.67901	3.85714	0.32834	12.06095
3.50000	0.45115	14.12500	3.31505	4.26087	0.21295	16.24199
4.00000	0.43496	18.50000	4.04687	4.57143	0.13876	21.06808
4.50000	0.42355	23.45833	4.87509	4.81188	0.09170	26.53866
5.00000	0.41523	29.00000	5.80000	5.00000	0.06172	32.65346
5.50000	0.40897	35.12500	6.82180	5.14894	0.04236	39.41235
6.00000	0.40416	41.83333	7.94059	5.26829	0.02965	46.81519
6.50000	0.40038	49.12500	9.15643	5.36508	0.02115	54.86198
7.00000	0.39736	57.00000	10.46939	5.44445	0.01535	63.55261
7.50000	0.39491	65.45834	11.87948	5.51020	0.01133	72.88713
8.00000	0.39289	74.50000	13.38672	5.56522	0.00849	82.86546
8.50000	0.39121	84.12500	14.99113	5.61165	0.00645	93.48763
9.00000	0.38980	94.33333	16.69273	5.65116	0.00496	104.75360
9.50000	0.38860	105.12500	18.49152	5.68504	0.00387	116.66340
10.00000	0.38758	116.50000	20.38750	5.71429	0.00304	129.21690

Table C.20 The Elements

Name	Symbol	Atomic Number	International Atomic Mass, 1966	Name	Symbol	Atomic Number	International Atomic Mass, 1966
Actinium	Ac	89	–	Curium	Cm	96	–
Aluminum	Al	13	26.9815	Dysprosium	Dy	66	162.50
Americium	Am	95	–	Einsteinium	Es	99	–
Antimony, stibium	–	–	–	Erbium	Er	68	167.26
–	Sb	51	121.75	Europium	Eu	63	151.96
Argon	Ar	18	39.948	Fermium	Fm	100	–
Arsenic	As	33	74.9216	Fluorine	F	9	18.9984
Astatine	At	85	–	Francium	Fr	87	–
Barium	Ba	56	137.34	Gadolinium	Gd	64	157.25
Berkelium	Bk	97	–	Gallium	Ga	31	69.72
Beryllium	Be	4	9.0122	Germanium	Ge	32	72.59
Bismuth	Bi	83	208.980	Gold, aurum	Au	79	196.967
Boron	B	5	10.811	Hafnium	Hf	72	178.49
Bromine	Br	35	79.904	Helium	He	2	4.0026
Cadmium	Cd	48	112.40	Holmium	Ho	67	164.930
Calcium	Ca	20	40.08	Hydrogen	H	1	1.00797
Californium	Cf	98	–	Indium	In	49	114.82
Carbon	C	6	12.01115	Iodine	I	53	126.9044
Cerium	Ce	58	140.12	Iridium	Ir	77	192.2
Cesium	Cs	55	132.905	Iron, ferrum	Fe	26	55.847
Chlorine	Cl	17	35.453	Krypton	Kr	36	83.80
Chromium	Cr	24	51.996	Lanthanum	La	57	138.91
Cobalt	Co	27	58.9332	Lawrencium	Lr	103	(257)
Columbium, see niobium	–	–	–	Lead, plumbum	Pb	82	207.19
–	–	–	–	Lithium	Li	3	6.939
Copper	Cu	29	63.546	Lutetium	Lu	71	174.97

Source: Reprinted by permission from Reynolds, W. C., Perkins, H. C., 1977. Engineering Thermodynamics, second ed. McGraw-Hill, New York, p. 653 (Table B-15).

Thermodynamic Charts

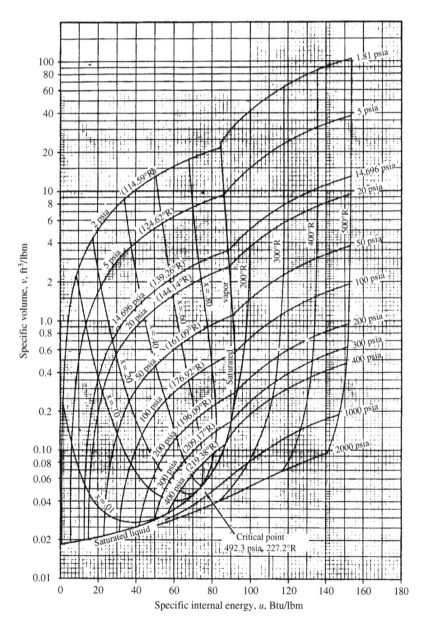

FIGURE D.1

v-u chart for nitrogen. Based on National Bureau of Standards TN-129A. *(Sources: Reprinted by permission from Reynolds, W. C., Perkins, H. C. Engineering Thermodynamics, second ed., 1977, McGraw-Hill, New York.)*

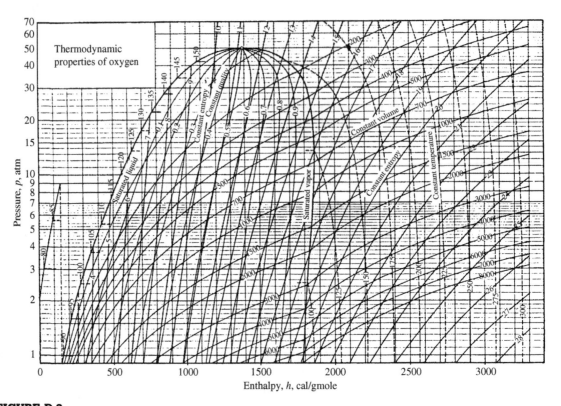

FIGURE D.2

p-h Chart for oxygen. From NBS D-2573. 1 cal = 4.184 joules, *v* in mL/gmole, *T* in K, *s* in cal/(gmole · K). *(Also reprinted by permission from Reynolds, W. C., Perkins, H. C. Engineering Thermodynamics, second ed. McGraw-Hill, New York.)*

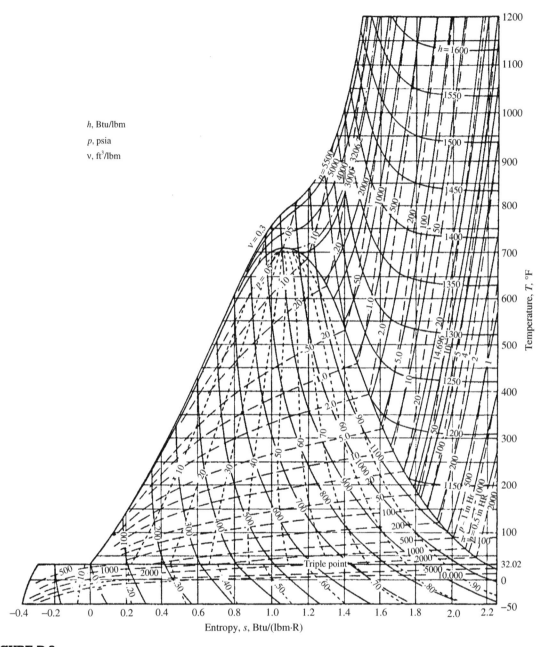

FIGURE D.3

T-s Chart for water. *(Sources: Reprinted from Keenan, J. H., Keyes, J., 1936. Thermodynamics Properties of Steam. Wiley, New York. Copyright 1936 John Wiley & Sons. Reprinted by permission of John Wiley & Sons, Inc. As adapted by Lay, J. E. Thermodynamics. Charles E. Merrill, Inc., Columbus, OH, 1963, by permission.)*

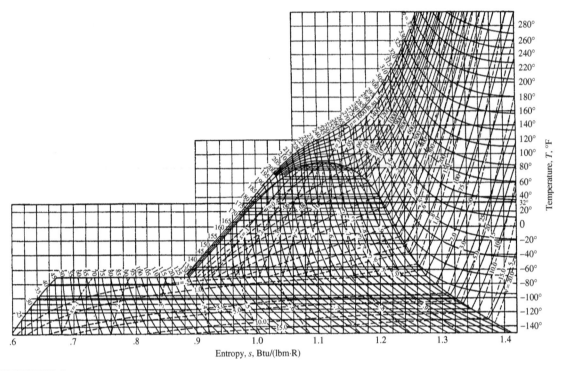

FIGURE D.4

T-s chart for carbon dioxide. Diagram copy supplied by General Dynamics Corporation, Liquid Carbonic Division *T* in °F, *h* in Btu/lbm, *v* in ft³/lbm, *s* in Btu/(lbm · R); at critical point *p* = 1066.3 psia *T* = 87.8°F. *(Sources: Reprinted by permission of General Dynamics Corporation, Liquid Carbonic Division. Also reprinted from Reynolds, W. C., Perkins, H. C. Engineering Thermodynamics, second ed., 1977, McGraw-Hill, New York.)*

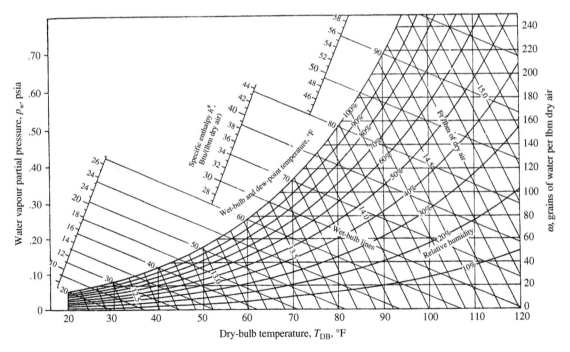

FIGURE D.5

Psychrometric chart for water (English units). Courtesy of the General Electric Company. Barometric pressure = 14.696 lbf/in.² (1 lbm = 7000 grains) *(Sources: Courtesy of the General Electric Company. Also reprinted from Reynolds, W. C., Perkins, H. C. Engineering Thermodynamics, second ed., 1977, McGraw-Hill, New York.)*

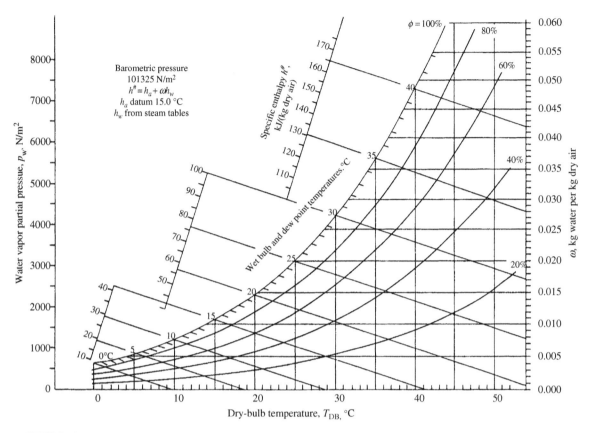

FIGURE D.6

Psychrometric chart for water (metric units). *(Source: Reprinted by permission from Reynolds, W. C., Perkins, H. C. Engineering Thermodynamics, second ed., 1977, McGraw-Hill, New York.)*

Printed and bound by CPI Group (UK) Ltd, Croydon, CR0 4YY

03/10/2024

01040319-0017